HAWKING INCORPORATED

HÉLÈNE MIALET

HAWKING INCORPORATED

Stephen Hawking and the Anthropology of the Knowing Subject

The University of Chicago Press Chicago and London

Hélène Mialet has held positions at Cornell, Oxford, and Harvard Universities. She currently lives and teaches in Berkeley, California.

The University of Chicago Press, Chicago 60637
The University of Chicago Press, Ltd., London

Printed in the United States of America

21 20 19 18 17 16 15 14 13 12 1 2 3 4 5

ISBN-13: 978-0-226-52226-5 (cloth)
ISBN-10: 0-226-52226-1 (cloth)
ISBN-13: 978-0-226-52228-9 (paper)
ISBN-10: 0-226-52228-8 (paper)

Library of Congress Cataloging-in-Publication Data

Mialet, Hélène.
Hawking incorporated: Stephen Hawking and the anthropology of the knowing subject / Helene Mialet.
pages; cm
Includes bibliographical references and index.
ISBN 978-0-226-52226-5 (cloth: alkaline paper)—ISBN 0-226-52226-1 (cloth: alkaline paper)—ISBN 978-0-226-52228-9 (paperback: alkaline paper)—ISBN 0-226-52228-8 (paperback: alkaline paper) 1. Hawking, S. W. (Stephen W.) 2. Physicists—Great Britain. 3. People with disabilities in science. 4. Communication in science. 5. Self-help devices for people with disabilities. 6. Mind and body. I. Title.
QC16.H33M53 2012
530.092—dc23 2011050728

♾ This paper meets the requirements of ANSI/NISO Z39.48-1992 (Permanence of Paper).

CONTENTS

ACKNOWLEDGMENTS

I have incurred many debts in writing this book, beginning with a Marie Curie Grant from the European Commission's Human Capital and Mobility Program that allowed me to conduct my initial research at Cambridge University. I am very grateful to the Department of History and Philosophy of Science at Cambridge for welcoming me and providing such a stimulating intellectual environment in which to launch this project. I was able to continue my research thanks to the support of the Maison Française of Oxford and the Museum for the History of Science of Oxford, the Max Planck Institute for the History of Science in Berlin, the Department of Science and Technology Studies at Cornell University, the Departments of Rhetoric, Anthropology, and Sociology at the University of California, Berkeley, and the Department of History of Science at Harvard University. Part of this research was also supported by grants from the Society for the Humanities at Cornell University and the President's Council of Cornell Women.

Many have contributed in one way or another to the making of this book. Indeed, this fact illustrates one of its principal arguments—that an individual is always a collective. I would like especially to thank Bruno Latour, Michael Lynch, Simon Schaffer, and Michael Wintroub for their wonderful inspiration, advice, and criticism. In addition, many friends, colleagues, and students have helped me in this long journey; I cannot thank them enough, but I would at least like to acknowledge my debts to Madeleine Ackrich, Ken Alder, Malcolm Ashmore, David Bates, Jim Bennett, Robin Boast, Robert Brain, Michael Bravo, Charlotte Cabasse, Jimina Canales, Nina Caputo, Florian Charvolin, Yves Cohen, Lawrence Cohen, Harry Collins, Olivier Darrigol, Lorraine Daston, Arnold Davidson, Richard Drayton, Nathalie Dubois-Stringfellow, Soraya de Chadarevian, Melanie Feakins, Marianne Ferme, Claudio Fogu, John Forrester, Marion Fourcade, Beate Fricke, Christophe

Galfard, Liza Gitelman, Ken Goldberg, Jan Golinski, Sudeshna Guha, Ian Hacking, Mitch Hart, Cori Hayden, Antoine Hennion, Anita Herle, Arne Hessenbruch, Don Idhe, Sheila Jasanoff, David Kaiser, Wulf Kansteiner, Devva Kasnitz, Lauren Kassell, Eivind Kahrs, Chris Kelty, Kevin Knox, Catherine Kudlick, Dominique LaCapra, Svante Lindqvist, Jean-Pierre Luminet, Lyle Massey, Andreas Mayer, Chandra Mukerji, Mary Murrel, Richard Noakes, Stefania Pandolfo, Vololona Rabeharisoa, Jessica Riskin, Margaret Rigaud, Gene Rochlin, Oliver Sacks, Natasha Schull, Sam Schweber, John Searle, Ann Secord, Jim Secord, Evan Selinger, Steven Shapin, Russel Shuttleworth, Otto Sibum, Peter Skafish, Isabelle Stengers, Lucy Suchman, Charis Thompson, Keith Topper, John Tresh, Heidi Voskuhl, Loïc Wacquant, Andrew Warwick, Hayden White, Mario Wimmer, and Alexei Yurkchak.

Aryn Martin helped me with the collection of information at the beginning of this project, Liz Libbrecht with translating some of my work, Laurie McLaughlin with the transcription of recordings, Susan Storch and Peter Skafish with the editing of the manuscript, and Erica Lee with the preparation of the manuscript. I thank all of them.

I have presented portions of this book at Cornell University, the University of Michigan, Northwestern University, MIT, Harvard, Stanford, the Universities of California at Berkeley, San Diego, and Davis, the University of British Columbia, Cambridge University, Oxford University, Manchester University, Brunel University, Cardiff University, Imperial College, the Max Planck Institute, the Centre de Sociologie de l'Innovation de l'Ecole Nationale Supérieure des Mines de Paris, and the Ecole des Hautes Etudes en Sciences Sociales. I would like to thank all the many participants in these talks, conferences, and seminars for their comments, feedback, criticisms, and interest.

In addition, I would like to thank all those I interviewed in my research for this book; though I did not cite everyone who took the time to answer my many questions, all played an important role in its completion. I thus want to thank each of the actors I interviewed, and especially Professor Hawking, for their time, consideration, and insight. Without them, this book would not have been possible.

Finally, my parents, my brother, and my family-in-law never despaired of seeing this book appear. I thank them for believing in it. Michael Wintroub, with his love for language and ideas, helped me at every step of this project; he and Maxime, with their unconditional support and love, were the driving force behind this project. I dedicate this book to them.

Portions of this book have been previously published. Chapter 5 appeared as "Reading Hawking's Presence: An Interview with a Self-Effacing

Man," in *Critical Inquiry* 29, no. 4 (2003): 571–98, and elements of chapter 4 were published in "Do Angels Have Bodies? Two Stories about Subjectivity in Science: The Cases of William X and Mr. H," in *Social Studies of Science* 29, no. 4 (1999): 551–82. I would like to thank these journals for their permission to reproduce my work here.

Introduction

By way of a prologue, I shall start with a thought experiment suggested by John Locke in his *Essay Concerning Human Understanding*.[1] The experiment is based on the following riddle: What would happen if, instead of eyes, scientists had microscopes in their eye sockets? The answer: Equipped with such prosthetic eyes, they would get to the essence of things, for they "would come nearer the discovery of the texture and motion of the minute parts of corporeal things, and in many of them probably get ideas of their internal constitutions," but they would simultaneously become angels, for then they would be "in a quite different world from other people: nothing would appear the same to ... [them] and others: the visible ideas of everything would be different." And Locke adds, "So that I doubt whether [they] and the rest of men could discourse concerning the objects of sight, or have any communication about colors, their appearances being so wholly different."[2] Hence, what they would gain in divinity would be lost in humanity, for humans would no longer be able to communicate with them. Thus Locke concludes,

> Since we have some reason ... to imagine, that spirits can assume to themselves bodies of different bulk, figure, and conformation of parts—whether one great advantage some of them have over us, may not lie in this, that they can so frame and shape to themselves organs of sensation or perception, as to suit them to their present design, and the circumstances of the object they would consider.... The supposition at least, that angels do sometimes assume bodies, needs not startle us; since some of the most ancient and most learned Fathers of the church seemed to believe that they had bodies: and this is certain, that their state and way of existence is unknown to us.[3]

Locke
Essay
Concerning
Human
Understanding
Good 115

By error or by chance, I think I have discovered an angel. He does not have microscopes for eyes, but he does have a synthesizer for a voice, and instead of his body's movements, he has a wheelchair and a computer. Stephen Hawking has undergone a series of trials in his life, starting in 1963 when, at the age of twenty-one, he developed amyotrophic lateral sclerosis (more commonly called Lou Gehrig's disease), characterized by muscular atrophy.[4] In 1985 he contracted pneumonia, underwent a tracheotomy, and consequently lost his voice definitively. In overcoming these ordeals, Stephen Hawking was to become an angel. Indeed, between Professor Hawking in his wheelchair and the universe there seem to be no mediators—or only one: his mind, as expressed in the following reports from the popular press.

> Mind over Matter: Stephen Hawking roams the cosmos from the confines of a wheelchair. *Telegraph Sunday Magazine*[5]

> Stephen Hawking Probes the Heart of Creation: His scientific genius soars from his severely crippled body—to unfold the deepest mysteries of the Universe. *Reader's Digest*[6]

> Roaming the Cosmos: Physicist STEPHEN HAWKING is confined to a wheelchair, a virtual prisoner in his own body, but his intellect carries him to the far reaches of the universe. *Time*[7]

> Reading God's Mind: Confined to his wheelchair, unable even to speak, physicist Stephen Hawking seeks the Grand Unification Theory that will explain the universe. *Newsweek*[8]

> Despite the adversities—or perhaps, as some have suggested, because of them—he would continue to climb. Where his feet could not go, his mind would soar. *Stephen Hawking: A Life in Science*[9]

Stephen Hawking incarnates the mythical figure of the lone genius.[10] He also represents "the perfect scientist" as we have come to imagine him or her in the rationalist Cartesian tradition. The man who can do nothing but "sit there and think about the mysteries of the Universe," this intellect, liberated from his body and seemingly emancipated from everything that clutters the mundane mind (such as emotions, values, and prejudices), is thus in a position to contemplate and understand the ultimate laws of the Universe.

Hawking has become an emblem for the ideology that dominates our understanding of science, namely, that science is practiced by disinterested

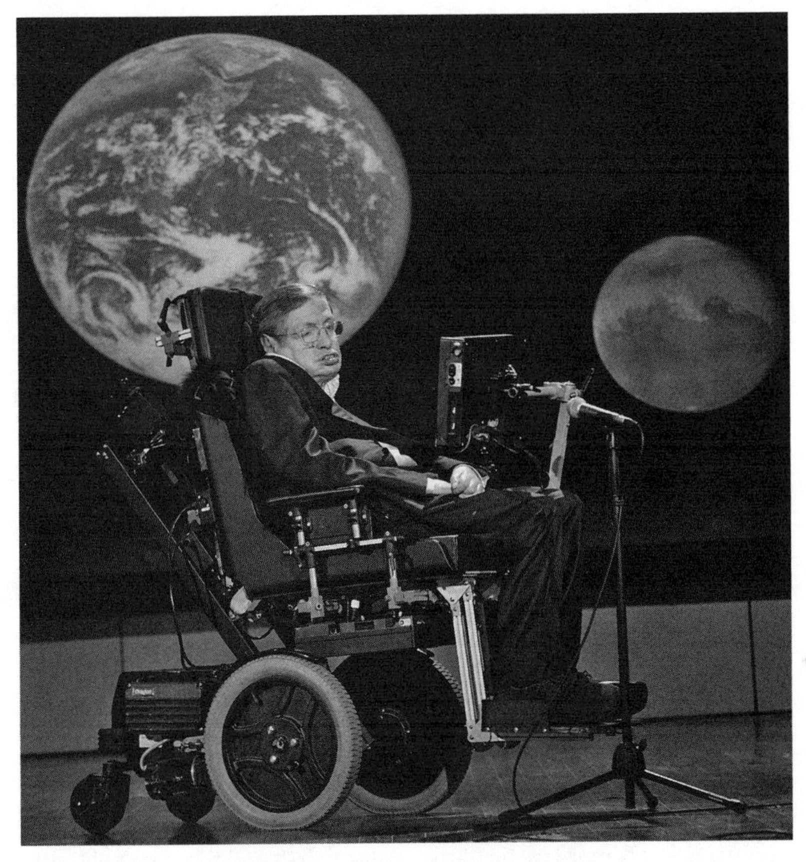

Figure 1. Hawking the angel. Photo courtesy of NASA/Paul E. Alers.

scientists who are able to transcend the political, social, and cultural spaces that their bodies inhabit in order to live in the unadulterated world of the pure mind. As the historian of science Steven Shapin points out, "Unlike the maker of handicrafts, the maker of universal knowledge is not tied to his workshop; unlike the hunter or fisherman, he is not dependent on the movement of his prey; unlike the professional, he need not await the solicitation of clients or patients; and unlike the athlete, his work is not conditional on the state of his body. The philosopher expresses ultimate freedom from the world's particular ties and demands. He is at home everywhere and nowhere. His disengagement from the social world is a symbolic voucher of his integrity."[11]

Historians, anthropologists, and sociologists of science have been trying to escape this figure for years. Indeed, if it seems difficult to draw together under the same banner these different disciplines, schools, and methodologies, one can at least localize points of agreement when they describe how knowledge functions. One common point is a constant opposition to the notion of solitude. To produce knowledge, we are told, is not a solitary affair; it is a collective—a social—enterprise. If the word *solitude* emerges in written accounts by and about scientists, it is simply a matter of concealment (deliberate or not) necessary for the validation of knowledge: the assistants are hidden (Shapin, Sibum), the conversations are effaced (Latour), the memory is reinvented, the mythical accounts circulate (Schaffer).[12] This collective production of knowledge is characteristic not only of the context of justification (Popper has already underlined its role in the test of falsification),[13] but of the context of discovery. What is at stake behind this questioning of solitude is the specificity of the cognitive operations of those who produce knowledge. Solitude forbids access to how knowledge is made; it forbids the presence of the other (the analyst or ethnographer); and, even if access were possible, it says that there is nothing to see, just an individual meditating silently, or a "thinker" translating ideas to a piece of paper. Where in this context are the material and social brought into play? What are the paper, pens, books, and colleagues doing? And where do ideas come from? This is the picture of the ascetic or solitary knowing subject that one has to deconstruct. To understand how science is really made, one has to enter forbidden territories and follow the procedures, techniques, and conversations that participate in the fabrication of scientific facts.[14] One has to follow the eyes, the hands, and the context of those who know. Only after doing this indispensable work will we be able to discover if scientists in general—and, more particularly, the formalists and theorists (who supposedly work with only their minds) or certain exceptional geniuses—think using a "method" or a "rationality" specific to science alone.

By trying to reincorporate thought in its (social-historical) environment and to exteriorize cognitive competencies in a network of human and nonhuman actors, we can call into question many of the dichotomies upon which the specificity of our knowledge has traditionally been thought to be based: the dichotomy between we Westerners, with our rational minds, and the others, the uncivilized *bricoleurs*; the dichotomy within our own culture between those who possess theoretical knowledge and those who don't;[15] the dichotomy within the scientific community between the big ones (the geniuses) and the small ones (the technicians); and, finally, the dichotomy between humans and nonhumans, men and machines. It is, according to the school of the sociology of science known as actor-network theory (ANT), difficult to impose clear distinction beforehand. Thus, we could say that the list of actors, or *actants*, to use the semiotic terminology, and the competencies that they possess are, in part, the core issue in the process of knowledge production.[16] And indeed this idea—of endowing objects with agency—remains among the most controversial hypotheses in contemporary social theory and science studies.[17]

The understanding of scientific practice as the redistribution of knowledge reveals a whole set of new and challenging questions. If, for example, knowledge is dislodged from its place of production—the mind or brain, depending on the school—the question of the origin of knowledge once again becomes enigmatic. When you say, "This is the person who first had the idea of" or "Einstein discovered general relativity," is this a myth or is it merely the fruit of a process of attribution? And what about psychology? To what extent can one exteriorize cognitive competencies normally limited to the heads of a few scientists, such as genius, creativity, expertise, and so on? Is displacing intelligence from the incorporeal mind to the knowing body just another form of mystification—not fundamentally different from that originating in the mythic genius of Einstein's brain so well described by Roland Barthes in his book *Mythologies*? Finally, with the hypothesis of the redistribution of knowledge, what was crystallized in an individual—the genius, the intellectual capacities, the ideas, the science—is opened up into the environment and ends up dissolving the individual's singularity by emptying her of her innate properties. Indeed, if scholars of science and technology studies have reintroduced a flesh-and-blood scientist into the process of knowledge production, she is a subject who is (part of) a "collective" body. Either subjectivity is immediately social, as in relativist sociology,[18] or, as in the case of actor-network theory, the subject becomes a spokesperson for an association of actors. Pasteur, as described by Latour, is not a body endowed with a mind, "or rather, he is far more than a body interacting with other bodies. He is a combination of a large number of elements that, through the links between them, produce Pasteur-the-great-researcher."[19] Thus, the definition of the human actor and her specificity once again becomes a central issue—indeed, what becomes of the individual in light of the collective bodies composed of humans and nonhumans, whether we call them actor-networks (to use Callon and Latour's terminology) or cyborgs (to use Donna Haraway's)?[20] These are some of the questions that I address in this book.

To do so, I have chosen to study Stephen Hawking. Stephen Hawking is intriguing because he seems to incarnate everything against which the discipline of science and technology studies constitutes itself: the possibility of producing scientific knowledge on the basis of his rational mind alone. He illustrates the extreme conditions of thought: solitude, pure mind, intelligence beyond reach, the force of reason, absolute genius. To those who might challenge this characterization and who argue that before we look for specific cognitive competencies in the minds of scientists such as Haw-

king, we must first describe the specificity of what they do practically, I ask how? "First, they say, we need to follow their eyes." To this, I answer: "This man can't move his head." "Follow his hands then." "Well, I can't do this either for he hasn't been able to use them for years." "His speech maybe." "How? He has lost his voice!" Yet, *this man* does *think, he* produces theories, *he* publishes, *he* gives conferences, and we say that *he* is a genius.[21] But, what does *he* really *do*?[22]

This question is important insofar as the knowing subject has been traditionally associated with the individual, disembodied, rational actor.[23] Indeed, the image of the brain-in-the-vat has become an extremely powerful part of our modern mythology. According to this myth, it is the "disincorporated mind" that makes us different and "superior." It has justified imperialism (the civilized versus the savage); it has justified hierarchies in our own societies (the expert versus the laypersons; the intellectual versus the workers); it has justified hierarchies in our laboratories (the scientist versus the technicians); it has justified ontological hierarchies (the human versus the nonhumans).[24] In reconceptualizing the knowing subject in ways that reconcile the decentering analyses of recent science and technology studies that attribute knowledge production to the social and material collective,[25] while still coming to grips with the singularity of individuals and events, I have tried to foster a new methodology and to propose new theoretical tools that can help us rethink these hierarchies and the myth of the individual, disembodied, rational actor that informs and justifies them.

Choosing Hawking as an extreme case, I want to see if—and to what extent—we can move beyond the notion of the self-sufficient genius. Based upon extensive ethnographic research and on a series of more than one hundred interviews with Hawking himself, his assistants, his nurses, his graduate students, his colleagues, physicists, journalists, filmmakers, archivists, artists, and designers of computer programs and hardware for the disabled, this book is an ethnographic study of "a knowing subject" that treats Hawking's unique situation as a privileged case by which to address larger questions having to do with genius, singularity, identity, subjectivity, corporeality (or the mind/body problem), distributed agency, socio-technical networks, scientific practice, formalism, language, cognition, creativity, expertise, and the frontiers of humanity. *Hawking Incorporated*, therefore, is not about Stephen Hawking—that is, it is not about a man, and it is not a biography.[26] It is a work of empirical philosophy. Moreover, though ethnographic descriptions of Stephen Hawking's life and work may allow us to grasp elements constitutive of his presence and the creation and maintenance of his identity, this book never claims to tell us who Hawking "really" is or to "capture" him fully. Rather, one of its principal themes is to question the very possibility of knowing who or where he is. This is reflected in the diversity of the material used and the structure of the book.

Hawking Incorporated is divided into seven chapters. Each chapter opens a door on a different collective at work whose description allows us to understand (the production of) an actor who performs, publishes, thinks, represents, converses, memorializes, and incarnates. The first chapter takes us from Professor Hawking's scientific laboratory—his office—to the public stage. I here provide a thick description of the network of competencies—the computer/the synthesizer/the personal assistant/the graduate assistant/the nurses—that transforms a man deprived of speech and movement into "the genius we all know." We thus follow the work of incorporation and redistribution of Hawking's competencies in his environment into the mise-en-scène of what I call his extended body and follow how Hawking's identity is created, maintained, and reproduced from one place to another. In the second chapter, we go back to the laboratory to describe another part of Hawking's extended body, the extensive human, material, and machine-based networks that enable him to work or to produce theories (i.e., students/colleagues/computer). Here, in following how Hawking's scientific papers are produced, I show how all the fundamental aspects of "pure science"—thought, proof, calculation, the contexts of discovery and of justification, and the reception of the scientist's work—are incorporated and distributed throughout the laboratory. Though the content of Hawking's work plays an important role in this book, I do not intend to provide a comprehensive account of it. Others have already done this.[27] Rather, based upon interviews as well as the extensive textual sources on his professional life, I provide an account of *his* way of working. In the third chapter, I follow the processes of attribution singularizing Hawking by taking my analysis a step further. Indeed, it is often said that "Hawking thinks geometrically." To understand what this means, I again reconstruct the network of competencies that allows him to do this—that is, to think by means of the "prosthetic" of diagrams. This will perhaps tell us something about the way in which certain theoreticians think (visual versus analytical), and about the way in which Hawking in particular does what he does. In the fourth chapter, we focus on the negotiations around—and the creation of—a BBC/Horizon documentary about Hawking's recent work. Here, I show how the media—together with the scientist himself—transforms this collective body into a disincorporated brain. Through a meticulous analysis of the press and interviews by and with scientific journalists, I pinpoint the mechanisms by which Hawking's status as a scientific genius is constituted (i.e., the standardization of autobiographical accounts, the media's representation of the scientist's body, the intervention of the scientist in the construction of his own myth, etc). We also see how the scientist himself—in his writings, in his relationships with his peers, and in his work (either intentionally or unintentionally)—uses his disabled body in such a way that he himself can be seen to participate in the collective construction of his identity as a perfect Cartesian subject, that is, as a mind without a body. We then follow how his writings and appearances are used, enriched, and recontextualized—by himself and others—so that, through this perpetual repetition and accumulation of written artifacts, a relatively standardized image of a human being endowed with stable physical, intellectual, and personal qualities emerges (e.g., his smile, his jokes, the role of his wife in his survival, his capacity to think in a geometrical way, etc). In the fifth chapter, we meet Professor Hawking in person for the first time and learn how a multiplicity of narratives "about" Hawking produce the stable identity that face-to-face contact disallows. The closer we get to him, the further away we seem. We no longer know "who" or "where" he is. All the categories we normally use in thinking about a person—a body, a machine, a mind, an interaction, a conversation, a text, and speech—become blurred. In his presence, Hawking seems more difficult to find than ever. This is where the confusion and fragmentation of identity is pursued by its recomposition. Hence, we follow how the machine used by Hawking articulates three different bodies whose nature I describe—a natural body collectivized, a collective body naturalized, and a sacred body as Lucasian Professor of Mathematics. In the sixth chapter, we see how Hawking's body becomes the locus of different interests, and how his body is distributed in his environment, and, by a reverse movement, how his presence and his singularity are constituted. Indeed, I show how certain groups, in attaching their own interests to (different versions of) Hawking's identity, are able to extend their presence in Cambridge, and, simultaneously, how Hawking, through this process, distributes his extended body in and beyond the city of Cambridge. To do this, I follow the creation of the Permanent Hawking Archive (supported by Intel's founder, Gordon Moore). Like Hawking's computer, a collective of librarians is now going through what "Hawking" has produced (and what has been produced about him) trying to classify, reorganize and preserve "his" writings (e.g., they have to decide what is unique or not, what is written or spoken, what is personal or not). Here, Hawking's reliance on the computer makes visible—and, indeed, is exemplary—of developing criteria regarding the storage and preservation (or not) of digital information. In the final chapter, we move to the most improbable scene: Hawking meeting HAWKING[28] the statue, an artwork produced to grace the entryway of the Department of Applied Mathematics and Theoretical Physics (DAMTP). In this chapter, I provide a detailed account of what took place during this meeting, where the presence of the statue allowed everybody (Hawking, his colleagues, the assistants, the sculptor) to compare—and interrogate the boundaries separating and constituting—"the real" Hawking and his "simulacra/copy/representation."

The methodological thread running throughout this book is constituted by the possibility of being in the presence of Hawking or not, of being close or far away from him. Here it is not only the scientist, but also the ethnographer who is equipped with a kind of adjustable microscope, to use Locke's metaphor, that allows us to zoom in and zoom out at will. Thus, in the first chapter, the physical presence of Hawking the man and his competencies are reconstructed from a series of interviews done with the people surrounding him, his assistants, but *without having access* to the man himself. In the second and the third chapters, we follow other beings who create him, constitute him, and extend him: for example, the students and the diagrams that make abstraction possible. In the fourth chapter, we examine accounts made about and by him and see—through the proliferation and repetition of texts and images diffused by the media—the qualities of Hawking begin to emerge. In the fifth chapter, we find ourselves *in the presence of* the man himself, while, simultaneously, the representation of the man disappears. We then follow how this man participates in the construction of his sacred body, to use Kantorowicz's formulation,[29] at the same time as he makes his flesh-and-blood body disappear. In the sixth chapter, we wander the streets of Cambridge to find his presence anchored in the architecture of the city and in the creation of a new library. We examine how this environment allows him to distribute his body through the traceability of the trajectory of "his" articles, books, and objects. In chapter 7, we observe him observing a representation of himself: a newly sculpted statue. In other words, over the course of the book, I try to describe the different materialities (the machines, the assistants, Hawking himself, the students, the colleagues, the diagrams, the journalists, the articles, the movies and the books, my own presence, the archivists, the artist, the architecture of the city, the archives, the statue) that constitute Stephen Hawking—that is, constitute his presence: make it durable, extend it, and conversely point back to him, which is to say, singularize him at the same time as they allow him to think, act, or be present. In this sense, insofar as this book is about individuality, it is also about the mediations that make *individual presence* possible. It is about what I call a distributed-centered subject.

1 The Assistants and the Machines

He who would fly through the hands and lips of men must long remain in his room. *Joachim du Bellay*

CAMBRIDGE, JULY 3, 1997
Slipping her head between the door and the wall, her body bending on one leg, one foot nonchalantly raised in the air, she asks him, "Do you want to give a conference at Oxford?"

Hawking raises an eyebrow.

His secretary makes an about-face and leaves. He said yes, she murmurs.

OXFORD, FEBRUARY 24, 1998
Eight hundred people are now waiting in the Oxford Town Hall. Tonight, thanks to the L'Chaim Society, Professor Stephen Hawking, "The Greatest scientist since Einstein, is going to speak on the 'Theory of Everything.'" The crowd is full to overflowing. It is eight o'clock. The curtain is still down and won't be going up. Tonight everything will be played in front of it.

LEARNING ABOUT HAWKING'S EXTENDED BODY

Here are two scenes—Hawking raising an eyebrow and HAWKING giving a talk; two spaces—Cambridge and Oxford; two times—July 3, 1997, and February 24, 1998; even two styles—a piece of daily life and a description of a public performance. Juxtaposed, these two vignettes seem to stand on their own. Nothing magic. Nothing mysterious. One pictures Hawking in his office; the other HAWKING on stage. Yet, between the two, an enormous amount of work, time, and organization is necessary to transform a man deprived of speech and movement into the genius we all know. This chapter draws on an extensive and in-depth series of interviews with Hawking's assistants to provide a thick description of the ensemble of human, material, and machine-based operations that give him the ability to act.[1] In this chapter, I first show that Stephen Hawking cannot act *without* a machine or without his assistants, who work day and night to maintain and stabilize his identity as Stephen Hawking the man and STEPHEN HAWKING the genius-physicist. Second, I examine the processes whereby this identity is replicated and reproduced as it moves from the laboratory to the public stage. As my description unfolds, I address questions related to language, communication, intentionality, action, identity, and presence.

Blurring Boundaries: Inside and Outside the Mind

WHAT DO THE MACHINES DO?

In 1963, Stephen Hawking developed amyotrophic lateral sclerosis, more commonly known as Lou Gehrig's disease. He is now totally paralyzed and confined to his wheelchair. In 1985 he definitively lost his voice from complications arising from pneumonia, which required him to have a tracheotomy. Thanks to a communication program called Living Center (designed and given to him by Walt Woltosz, of Words Plus, Inc.), and a speech synthesizer (designed and given to him by Speech Plus, Inc.), Hawking can communicate, write, and read. He is able to operate this entire system on his own by means of a simple switch (the commutator) that someone places in his hand and around which he curls his fingers.[2] The device consists of a highly sensitive on-off switch, which operates a cursor. Since the cursor automatically moves up and down, Professor Hawking can select words by pressing on the switch. When he has completed what he wants to say, he can send it to a speech synthesizer. The synthesizer and a small IBM-compatible personal computer (donated by David Mason of Cambridge Adaptive Communication) have both been mounted on his wheelchair. They run on batteries under the chair, although the computer also has an internal backup battery that will last for up to an hour, if necessary. The computer screen is mounted on the arm of his wheelchair.[3] Hawking can either speak what he has written, or save it to disk. He can then print it out, or call it back and speak it sentence by sentence. As he said, "Using this system, I have written a book and dozens of scientific papers. I have also given many scientific and popular talks."[4] According to his home page, he can also use Windows 95 through an interface called EZ Keys, again made by Words Plus. This program allows him to control the mouse with a switch through his selections shown on his desktop. He can also write using menus similar to those found in a program called Equalizer.[5]

Hawking has lost the ability to speak. What he "says," however, can still be heard through the medium of writing.[6] An able-bodied individual using a computer can run his fingers on the keyboard or move the mouse (with either his fingers or his hand) while the screen remains static. In Hawking's case, however, the body stays static while the screen unfolds before his eyes; he can stop this movement and select words with a click. More specifically, to write a word on his computer, Hawking has to follow four different steps:

he chooses a letter appearing in the first or the second half of his screen, then the half-screen chosen displays a series of words starting with this letter, he then chooses the relevant row, column, and, finally, the word on which he clicks. Hawking has become a very experienced and skilled user. As Woltosz, the inventor of these programs says, "Hawking is able to do ten steps per second—it's a blur to me!"[7] He is fast, indeed, but Equalizer and EZ Keys are also equipped with numerous visual and mechanical strategies that allow their users to save time. For example, the bottom of the Equalizer display shows the thirty-six most frequently used words in English (e.g., *I, to, the, and, but*, etc.). The most frequently used word is on the top of the first row (interestingly, this is the word *I*), and the second most frequently used word appears one line below on the right. Thus, the most frequently used words require less work than those used less often.[8]

Equalizer and EZ Keys also offer numerous other strategies to accelerate the writing process, such as "Word Completion," "Word Prediction," and "Next Word Prediction." For example, if one wants the word *infinite*, one chooses the letter *i*. The word display will automatically change to the six most frequently used words that begin with the letter *i*. If the word *infinite* does not appear, one chooses the letter *n*, and the six most used words beginning with letters *in* will show up. If *infinite* is on the list, one chooses the right number (4, for example), and the word *infinite* will appear on the screen. Then the program will switch to the second kind of prediction, called "Next Word Prediction," and will look up the six most recently used words by its user after the word *infinite*. If the next word, *space* for example, is one of the six shown, then one can simply choose the number of the word. Then the next six most frequently used words after the word *space* will appear.[9] As Woltosz says, "It's . . . a bootstrapping process, and very many times, when you create a sentence, you never complete a word, because the word prediction gives it to you. And very often you don't even have to begin to spell the next word, because it's already there; you simply select the whole word."[10]The program learns the word patterns of its user: "Simply select the correct word, and EZ Keys types it for you."[11] Put another way, "the program is trying *to guess* what word [the user] wants next."[12]

Equalizer also has "Word Prediction" and "Next Word Prediction," but in a less powerful form than EZ Keys. The presentation is different, and the vocabulary is smaller. For purposes of memory, the words appear horizontally on the screen. And because it is extremely difficult to check words alphabetically when they are displayed horizontally, their order of appearance is linked to their frequency of use or their most recent order. However, as Woltosz says, "We are much more efficient at scanning lists in alphabetical order than we are in frequency order.... If your telephone book had all the phone numbers that you used in frequency-of-use order, ... finding a phone number would be very unpleasant."[13] EZ Keys, thanks to its vertical and alphabetical presentation and its six rows of words (three rows, a space, and three rows [e.g., top, middle, bottom]) instead of four equally distributed rows as with Equalizer, allows for a quicker selection. EZ Keys was explicitly designed to exploit the fact that our eyes and brains deal well with groups of three. It is less demanding from a visual and cognitive point of view.

Both programs allow for the storage of routinely used sentences, comments, and exclamations so that they can be retrieved very quickly in everyday conversations. Equalizer has a multiple-page list of a small number of limited phrases, such as: "Sorry, what did you say?" "Thank you very much." "Can I have the bottle, please?" "Can I have suction, please!" etc. EZ Keys's Instant Phrases feature, on the other hand, has seven avenues of selection: F1—Closings, F2—Conversation, F3—Exclamations, F4—Greetings, F5—Insults, F6—Questions about you, and F7—Responses.[14] Hundreds of thousands of phrases and sentences can also be stored and categorized according to subjects such as family matters, sports, personal needs, food, jokes, or hobbies and then retrieved instantly.[15]

According to Woltosz, the way this device functions when it triggers a familiar topic is not fundamentally different from the way an individual talks. In this sense, the mechanization, standardization, and repetition of themes and phrases is linked less to the way the machine works than to the way language works. The machine records:

> If we got on a subject that I talk about frequently, ... if I was a sports fan, let's say, and I had a particular team that I liked, ... well, any of us who can speak if we got on a topic like that, we tend to say the same things over and over. There is a certain player that we like, ... such as X. "Oh! Did you see the play he made against Montreal," and ... "What a great move he made on this other player," ... so, you might use that sentence over and over.... With the ALS patient, they can store that ... store all their comments about their favorite team, and whenever they need them, they can pull them up and very quickly say them.[16]

The principal difficulty for an ALS patient occurs when he has to speak about something new. Deprived of an already constituted repertoire, he is obliged to construct a sentence from scratch. The main problem is speed: "No matter what we do, it is still too slow. When we speak, we tend to speak at around 150–200 words a minute.... The best we have seen would be someone like Stephen Hawking, who scans at a very, very fast rate, and at top speed he is probably going about 20 words a minute."[17]

EZ Keys also has a function called "Abbreviation Expansion." This allows the user to utilize abbreviations for frequently used words and phrases and for instant speech.[18] Indeed, what is characteristic about the way we communicate by voice is interruption. In carrying on a conversation, we are often interrupted by our interlocutors, or we talk while we are doing something else, like checking e-mail. This strategy enables communication across different platforms:

> While I am building a sentence, . . . suppose I want to say, "I want to buy a CD" or something. . . . I start to make a sentence, "I" "Want" . . . and about this time, someone walks into the room, and I want to say, "Hello, how are you" in the middle of my sentence, I choose "HH": "*Hello, how are you?*" (voice), [and then] it erases that and puts me right back where I was. That's what we call instant speech abbreviation. So, we can abbreviate things, and we can tell the program whether the abbreviation is just something that I want to type here or whether it is something that I want to say instantly. Typical instant speech would be like "GM" for "Good morning."[19]

Through this feature, the computer can also perform a number of different tasks, such as answering the telephone or printing a file. EZ Keys also makes possible a variety of activities in and out of the home. Using a modem or fax/modem, one can send and receive messages or explore the Web.[20]

Hawking is probably the last and only user of Equalizer today. This program has changed very little since 1985 when he first began to use it. It is much more limited than EZ Keys, having less memory (a sixteen-bit architecture) and therefore a much more limited vocabulary. Concerning its presentation, Equalizer uses a full-screen display, while EZ Keys, because it runs with other software, occupies only a small portion of the screen, usually the upper right corner. As Woltosz emphasizes, "Equalizer is a very primitive-looking program, because it runs in a DOS Window, so it has the old-fashioned DOS interface, with large characters and so on. It doesn't look at all like today's software would be expected to look."[21] EZ Keys, by contrast, is an all-in-one integrated computer/access system; the user doesn't have to stop what he is doing to have voice communication; lastly, EZ Keys offers many more strategies (time-saving features) to retrieve words. For all these reasons, Walt Woltosz comments,

> No one else uses Equalizer. . . . When [Hawking] asked for a change recently to Equalizer, my engineering manager had to hunt and hunt to try to find

> a computer that still had the file on it, with the source code, because all we had was a disk with a compiled version. . . . No one has used it for years and years. Only Hawking, again, because it's terribly inefficient compared to EZ Keys. [. . .] In its day, it was by far the most efficient program that was out, but we've learned a lot since that was written in the early 1980s. [. . .] By today's standards, it's a dinosaur."[22]

Nevertheless, in spite of Equalizer's limitations and the pressure from his environment to change programs and adopt EZ Keys for everything, Hawking has always refused to separate himself from Equalizer. As Woltosz said to me, "He believes he is faster [with Equalizer]; probably a better description is *that it is so burned into his brain* that he feels more comfortable [with it]. But I can tell you from direct observation that he is much slower with [it]—his wife and previous student assistants agree. [. . .] But you can't tell Stephen Hawking that; *he refuses to acknowledge it*."[23] "I'm sure," Woltosz said, "you can sense my frustration that Stephen is so adamant about using Equalizer. It is frustrating because I know it reduces his productivity, and it is frustrating because people see him communicating so slowly and attribute it to Words Plus—yet it is far from our state-of-the-art systems."[24] But, "obviously, we want him to be happy, so we accommodate him."[25]

Several years previously, over dinner, Andy Grove, the chairman of Intel, asked Hawking what kind of microprocessor he had in his computer. Hawking, Woltosz recalls, didn't know. Two engineers from Intel discovered that there was no Intel processor inside. "Well, that will never do," said Grove. He immediately contacted Woltosz and explained that he wanted to build two new computers specifically for Hawking using the latest technology. This way, he could use the most recent version of Windows with all the functionality it offers. This meant that Woltosz had to create a version of Equalizer for Windows:

> He had been on a DOS system up until then. And we hadn't done that. That was not a simple task. There were certain things that we had to do just to be able to read the switch . . . to tell when he was hitting his switch accurately. We were having a great deal of difficulty doing [this]. So, Intel assigned a small team of engineers to help us, and between what their engineers were able to do and what ours were able to do, we did figure out a way, finally, to make all this work in the Windows environment. But it was actually Hawking's requirement that was pushing us to get into the Windows system quickly and move away from DOS. That was probably the late '80s or 1989/1990. It is hard to remember now. [. . .] I think now he has upgraded again, and he has a newer machine. You know the technology keeps mov-

> ing ahead, and so what they might have made five years ago would be obsolete today. He is such an intense user that he would want as much power as he could get in the system in terms of how fast it runs and what different things he can do with it.[26]

Stephen Hawking has been using this system intensely for more than fifteen years: "It is really an extension of himself now," says Woltosz. "So if there is something that he doesn't like, he will tell us: I wish you could make it do this or that."[27] Apparently Hawking has been a gold mine of ideas: "He has made many good suggestions over the years, . . . probably because more than anyone else, he is a very expert user."[28] His suggestions are sent directly by e-mail to Woltosz or to the person in charge of users at Words Plus, or he or his assistants will get in touch with the distributor in England. We see, then, how the constant maintenance and modification of Hawking's computer system pushes companies that support him to innovate. Though they often give him what he wants as publicity for their products, his engagement with or resistance to the process of innovation plays an important role in the construction of his identity. In some sense, it is "his" identity.

Today (Wednesday, November 27, 2002), Hawking is quite anxious to switch to Windows XP, but right now he is not able to make the switch. If he wants to have Equalizer, he has to stay in Windows 95 or 98 or Millennium. Woltosz once again will have to completely rewrite Equalizer from scratch, in a new language, that will run on Windows XP.

> And that's a very big task to do for only one person. . . . Right now we're trying to decide. Intel has offered to do it, or they've offered to pay us to do it. I don't think anyone at Intel, anyone that we've talked to, has the background to really do it, very efficiently. . . . We've dealt with a number of Intel engineers over the years and, you know, tried to work with them, but they're just not given this particular assignment for a long enough time to become expert at it. It's usually a temporary thing, and then they move on to someplace else. And it's usually not their full-time job; it's just kind of a side issue."[29]

Stephen Hawking's voice has also been modified over the years, but these changes have been a significant challenge, because his voice is a 1986 voice. It came from a company called Speech Plus, which no longer exists. It was closely related to a voice synthesizer called the Dec Talk, which was developed by Digital Equipment Corporation (DEC). It was always and still is considered to be the highest quality "voice" available. The voices produced by these two companies sound nearly the same. "For most of us, if we heard the adult male voice from the two, we probably would not be able to distinguish them, and that is the voice that Stephen uses. Stephen can tell though," says Woltosz.[30] He recalls that when Hawking came and gave a lecture at Berkeley in 1987, Speech Plus was still in business. Their engineers came over to Berkeley and told Hawking that they had a new set of chips that improved the voice, and they wanted to put them in for him at no charge. They wanted to upgrade his voice. "So, they put in the chips, and he started using it, and he said 'that is not my voice. . . . *I want my voice back*.' So, he really identifies with that voice; it is his oral identity. Of course, it is getting to be pretty old technology now."[31] Several British companies have offered to provide a British voice synthesizer with a British accent. But in the same way that he is attached to the antiquated (DOS) look of his computer, he has always turned down even the slightest variation of his American male voice. Again as Woltosz says, "He really likes some very specific things that you and I would probably find hard to even distinguish between the voice he has and a speech box. . . .[32] When he gives a speech, . . . he apologizes for his American accent, which was always part of his opening jokes. And so he apologizes for it, but he doesn't seem to want to give it up."[33]

There is also an explicit link between the clarity of discourse, of an enunciation, and the supposed rationality of the genius. As Woltosz observes, "Stephen is one of these people who knows how to use the English language extremely well and doesn't want to talk to you in what we call 'telegraphic' speech, where you leave [out] a lot of words and you sound like a 1950s TV show probably with the American Indian: . . . 'Me want food.'"[34] Hawking also makes this link: "One's voice is very important. If you have a slurred voice, people are likely to treat you as mentally deficient: 'Does he take sugar?' This synthesizer is by far the best I have heard, because it varies the intonation, and doesn't speak like a Dalek. The only trouble is that it gives an American accent. However, the company is working on a British version."[35] Indeed, though he could have a brand-new voice, one that would be more naturally human, expressive, and easier to understand, perhaps even one with a British accent, the world would no longer think, in the words of his assistant, "Wow, that's Stephen Hawking, they would think, oh, that's my voicemail." Paradoxically, the American accent makes this mediation even more visible.[36]

Now, Intel is taking the characteristics of Stephen's hardware voice, which is a printed circuit board, and is transforming it into software for him. Again, if he wants to go to Windows XP, for example, in the newer hardware, it would be very difficult to accommodate a printed circuit board that is more than fifteen years old. The interface is quite cumbersome, and the printed circuit board comes from a company that doesn't exist anymore. As Woltosz says, "He has two [voices], he has a backup, but if that thing breaks, he's in trouble. He wouldn't have his voice. So, converting it to software makes everything much easier. It takes less power, less space and so on, and it's very easy to just move it from one machine to the next, when he upgrades his machine."[37] It also means rewriting EZ Keys and Equalizer, that is, rewriting parts of their software so that they work with the new voice. The goal is to make the system "younger" and easier to use. Hawking will get this voice soon. It was made just for him (it doesn't have any commercial potential); he will thus be able to maintain his identity. We see here the co-production of the intelligent system and its user.[38]

The versatility of these computer programs has also been extended to Hawking's living environment. Hawking has the most wired home imaginable. According to Woltosz, "Most of our ALS patients have a more limited environmental control that is a infrared control. We call it a U-control."[39] The optional U-Control II system allows the user to turn on a fan, play a favorite CD, or dim the lights. But again Hawking didn't have the ability to use this type of environmental control with Equalizer. Words Plus had to modify the program so that he could use another program to send signals out to a wireless transmitter to activate these different motors and controls.

> Now when Stephen wants his nurse—and his nurse is usually in her room in the back of the house—when he wants her to come where he is, he sends an environmental control signal that makes her door open. So that is his signal to her, and when she sees her door open, she knows that [Stephen needs her]. . . . He built a new home a few years ago, and his entire house is all set up with all these actuators. It is quite remarkable. [. . .] He is extremely independent, considering the level of physical disability he has.[40]

Hawking uses EZ Keys to surf on the Internet, to use Microsoft Office, Word, Excel, to play solitaire, to do his e-mail, to write technical papers, or to adjust/modify his home environment. He prefers to use Equalizer when he wants to relax, communicate, or have a conversation.

Obviously, Hawking can do exactly the same things on both his wheelchair and desktop computers. The only difference is that the desktops are connected to a network rather than a modem, and he can thus use e-mail and the Web much more quickly.[41] Before Hawking lost the use of one of his hands for switch action, he could run two switches at the same time: then, only really for his own personal convenience, he would use one computer, usually the wheelchair computer, to talk with someone, at the same time that he might be working, perhaps on a word processor, on his fixed computer. Now, because he can use only one hand, he has to take a few extra steps to switch back and forth between the word processing and the speech application. In late 2000, he began to lose the strength of his only working hand. Hawking thus started using a (Infrared/Sound/Touch) switch developed by Woltosz. By making only slight movement of his cheek, he can still activate the switch.[42]

WHAT DO THE ASSISTANTS DO?

Transforming Hawking into a Yes (or into a No)

Communication with Stephen Hawking involves reading signs on the computer when you don't know him, and signs expressed in body language (movement of his eyebrows, his mouth), when you do. This is how his assistants interact with him. One can see a strange symmetry between, on the one hand, a choice of responses proposed by the assistants and Hawking responding by yes or no and, on the other, the binary logic of his computer, which proposes a choice of sentences and words upon which he clicks. One also notices an extraordinary similarity in the interaction between Hawking and his technicians and Hawking and his computer: both seem to have been trained or programmed to operate in the same way. Each gives Stephen Hawking a series of options from which he chooses, after which the technician, nurse, or computer can perform it. Hawking's assistant's operation is a duplicate of the computer's, except that instead of responding to the subtle movement of Stephen's fingers, the assistant responds to the subtle movement of his eyes.[43]

"With Stephen," his graduate assistant says,

> you tend to have . . one-way conversations. You tend to phrase a question so the answer is yes or no. Which is actually an acquired art, I mean, it's something you can pick up. . . . So what he does is, . . . he raises his eyebrows for yes and not for no. And he's got an "I'm not very happy" [look] for, you know, . . . but the point is you can have a relatively quick, full conversation with Stephen without him actually having to use his computer to say anything, *as long as you phrase your questions correctly*. And that's something you pick up, and . . . you can have a whole conversation with Stephen without saying anything, I have! . . . You just base it on the fact that you know each other, that you know what each other is thinking, to an extent, and you've been working together, for a while. I mean, . . . you

> can have conversations with Stephen in which very, very little is spoken, but you both actually know what the other wants.[44]

This is also how Hawking's nurses communicate with him. Again, they ask him a series of questions like: "Do you want me to change hands?" "Do you want me to lift you up?" "Do you want to go?" And he will say yes or no. And if after having exhausted all the questions that come to mind they still don't know what "he says," and, for one reason or another, he is unable to use his computer, they will use spelling cards with all the letters of the alphabet. In general, they won't need more than three or four letters to know what he wants. "The nurses will know enough so he'd still be able to say, 'This is what I want.'"[45]

For a "one-way" conversation to become a "full" conversation, Hawking's assistants, as in any face-to-face interaction—but magnified with Hawking—anticipate his responses, complete his sentences, and sometimes give him a voice.[46] They operate as the computer does when it multiplies options to help its user construct sentences more quickly. To a certain extent, one can say that they have learned the patterns of their user. Like the computer, they also make mistakes. Hawking's frustration makes this evident; that is, sometimes he wants to say something else. There is a difference, however: more than anticipating and completing his sentences, Hawking's assistants project and attribute competencies to Hawking that are, for the most part, the product of their own labor. Hawking's role, in this regard, can be compared to the role of the psychoanalyst in the psychoanalytic setting. Similarly, Harold Garfinkel's discussion about an experimenter, falsely represented as a student counselor in training, is relevant here; thus, the analyst responds to his patients' questions with a predetermined sequence of yes and no answers without taking into account their content. The patients do almost all the work of the counselor (like the people around Hawking), but nevertheless attribute all the merit of their therapy to the counselor himself.[47]

Most of the time Hawking doesn't need his artificial voice to communicate, especially when he works. As his assistant recalls,

> The classic was once I actually had put the computer back in his chair, in the morning, and I had connected up the voice wrong—I had put in the wrong plug. And, it wasn't working. And . . . he left from work about four hours later, and that was the first time we realized that his voice wasn't connected. Because he never used it all day (laughs). You know, because he'd been sitting at his desk, using his desk computer's voice, not actually

> his chair computer. Or he'd been, you know, not been talking at all. So . . . he didn't notice until he left the building, and he wanted to say "Goodbye" and he couldn't say anything! (laughs)[48]

This little excerpt reminds us that gestures often communicate as well as—and sometimes even better than—words. Moreover, though speech is typically considered the primary means of communication for the severely disabled, like most professors or managers, Hawking spends more time writing than "speaking." "How do we spend our time?" says Woltosz. "I spend a lot of my day reading and writing, whether it's e-mail or documents for work, and we have to recognize that reading and writing are as much a part of communication as speaking. [. . .] The fact that you have a voice doesn't mean that you want to be talking every minute."[49]

Transforming a Yes into Hawking

David Goode, in his book *A World without Words*, considers how children who are totally deprived of formal symbolic language are able to communicate with their bodies when they are caught in a web of competencies organized around and with them. Hawking is endowed with formal language but is deprived of a voice and also partly (at least, to the untrained eye) of the gestural language to express it. As we have seen, most of the time, because there is no alternative, Hawking is transformed by his assistants into a yes or no who responds to a proposed menu.[50] At the same time, his special conditions impose the mechanization—the hierarchization, standardization, and routinization—of his human/machine-based environment.[51] In other words, for a yes or a no from Hawking to become performative, his assistants must be able to interact with him.[52] To do this, they need to learn certain tricks and skills that can only be developed through time and familiarity. On the other hand, they must also be organized in such a way that they can act on this yes or no so that Hawking, in his turn, can travel, talk, think, and perform. Depending on the particular network of competencies in which it is situated, a yes or no can have many different meanings. In this sense, language is as much a property of this collective as of the individual.

Thus, contrary to the solitary genius depicted by the media, Hawking resembles a manager at the head of a company, a company that has explicitly become his extended body. Walt Woltosz confirms this intuition: "Hawking," he says, "is probably the ultimate manager because everything *physically* that gets done, he has to do through other people, which is the definition of management."[53] Or again: "Well, management is largely about decision making. And . . . it doesn't take as much verbalization as it does thought, when you're making decisions. So, I mean, you might go through hours of considering different things to just say yes or no."[54]

Blurring Boundaries: The Private and the Public Self

THE SECRETARY (OR PERSONAL ASSISTANT): TRANSLATING THE OUTSIDE WORLD

The secretary of the Relativity Group of the Department of Applied Mathematics and Theoretical Physics is also Hawking's personal assistant (PA). She is in charge of all the requests for and about Hawking. She is also responsible for the recruitment of a team of nurses.[55] Like the other assistants (nurses and graduate assistants), she has, with time, developed a peculiar way of interacting with Hawking. She does not wait for him to speak—he usually doesn't say anything, and he doesn't need to—for she knows his answers by merely looking at his eyes, as one does with a child before it learns to talk. This is not a particularly difficult exercise, she tells me, as she has four children of her own.[56] It is also her job to deal with the "outside" world. On the day of my visit, for example, she had to show him an invitation received from an eminent professor to attend a conference in Chile. She knows that there is little chance of his going, but "He's the boss," and "I need to make him feel like he's the boss."[57] But what if Professor Hawking were to answer yes (with his eyes)? She knows that this simple word or sign would probably give her eighteen months of work organizing everything, that is, transforming a single word into Professor Hawking in person in his wheelchair talking through his synthesizer to a crowd of wildly enthusiastic Chilean scientists.[58] As she puts it to me, "Indeed, if this invitation arrived to somebody else, to you or to me, we could say, 'Oh yes, I could hop on the plane next week.' And we'd catch a bus to the airport, and we'd jump on a plane."[59]

But Hawking obviously can't do this. If he has to go somewhere, she will not be allowed to overlook the slightest practical detail, either concerning his journey and all the complications that a wheelchair equipped with a computer system implies as regards security, or concerning the financing of the trip and, above all, its scientific side. She will have to contact all the eminent persons that he will be meeting and know what questions they plan to ask him so that he can prepare the answers in advance. The aim will be to minimize, as far as possible, all unexpected interventions. Thus, in a certain sense, Stephen Hawking will be "preprogrammed" for all the formal meetings he will have.[60] "Every single last detail, as well as the title of the lecture," she says. "And if he's going to meet eminent people when he's there, I mean even if he needs to say, 'Hello. How very nice to be here. Thank you for inviting me.' He still needs to prepare that, so that he can say it. So every minute detail I need to sort out beforehand. And that's the large part of what I do."[61]

In addition to arranging his travels, she is also responsible for storing the dozens of letters received daily from the professor's admirers since his book *A Brief History of Time* was published; letters that he never sees and is not even aware of, for reading them would take every waking minute of his every day. She is also in charge of relations with the media. For example, she supplied John Gribbin and Michael White with all the personal information they needed to write a popular book on Hawking's life and achievements. Indeed, they never actually met Hawking in person while writing *Stephen Hawking: A Life in Science*. Thus, though she interacts with and fields requests from the media, she also needs to keep "such interests at appropriate levels."

Most of the time when people want him to appear on a television show, the response will be no. However, if his response is yes, he can't just take a train and go to London for the day. In this context, she will bring the world to Hawking, but this time "in person"; and instead of just "preprogramming" Hawking, she will also "preprogram" the television show. The television crew will come to meet him in the department, or she will find a place close to the department accessible for someone who is disabled. "He will only have a very small space of time in which it can happen. They need to be very focused."[62] The power relationship is thus inverted. Usually, the person interviewed can be filmed for hours and will appear for only two minutes. But for Hawking, this is impossible. He must prepare in advance. The secretary makes sure that the journalists know exactly what they want from him so that he can be perfectly and immediately integrated in the place they choose for him to appear. "Because it's going to take him days and days just to answer three or four questions. So if he does do an interview, especially for television, that is quite time-consuming. Again, *because it's detailed*."[63] Interestingly, this expression keeps pointing to the work of the collective that enables Hawking to act, while simultaneously annulling or minimizing its role.

But what she does like about her job is that she does it on her own, and no one tells her what to do or how to do it: "This is because he's a typical academic, and so he doesn't want to know about a lot of administrative stuff. And, because he's disabled and so obviously can't do stuff that people usually do, and he trusts me to do whatever needs doing, basically. So I have an enormous amount of freedom in actually doing the job, as I want to do it.

So *I'm the one who's prioritizing and deciding*, and . . . I enjoy the freedom, and the responsibility, and the challenge of trying to do something which lots of people think is impossible."[64] And what does she not like about it? "He is very difficult. He is very demanding, very egotistical, and . . . [pointing to my tape recorder she says] can we turn this off?"[65]

To a certain extent, one can say that Hawking's handicap makes apparent the normally hidden practices of academics who shift the administrative burden of their positions onto someone else and also the practices of managers who delegate competencies to their subalterns. As David, a young man who is quadriplegic that I interviewed, says,

> There are a lot of guys who operate the way I do, which is to tell their secretary, "I need a cup of coffee." "Grab me these books from the library," et cetera. They are not disabled. At some point in time when I needed help, and I was getting "What would you like"? I was like, "Bring me some coffee. I would like some toast." I realized that a lot of very wealthy people still don't have the kind of care that I get. *You have to be a king to have people wait on you hand and foot.* In some sense, if you look at it from that perspective, you are having all your needs taken care of by other people. You can feel luxurious. It is nice to be able to say, "I would like a cup of coffee," or "I would like to eat in bed tonight." "Scrub my back," or whatever else.[66]

In Hawking's case, there is a supplementary degree to this process of delegation. His own body is taken care of (like a king), but also all the boundaries between private and public are blurred. His competencies are more distributed than anyone else's. Though Hawking's secretary, as I have shown, sorts and arranges data according to his interests and to what he can deal with, she also takes charge of his private life. For example, she takes responsibility for his personal finances and legal problems (by the latter she probably means conjugal problems, implying his divorce and remarriage to his nurse), as well as personal matters such as paying his son's school fees, arranging for him to go abroad, taking him to the park, or simply fetching him from school.

THE GRADUATE ASSISTANT: ENGINEERING HAWKING'S TRANSLATION

This collective of humans and machines to which Hawking is attached is what I call his extended body. It keeps modifying itself in harmony with the modification of Hawking's flesh-and-blood body and his fame. The computer has modified Hawking's environment and keeps being updated.

Through it, he can "talk" and "write," thus allowing his students to quit their jobs as assistants or nurses and become students again. The nursing would now be for the nurses; thus four were hired to work around the clock. The department secretary became his personal assistant, doing mainly administration and management. Then, unable to deal alone with the immense workload created by Hawking's fame, especially after the publication of *A Brief History of Time*, she was able to recruit a graduate assistant (GA) financed by Cambridge University to "help the Professor in all areas in which he has difficulty due to his handicap," for example, taking care of his travel arrangements.[67]

A background in physics, mathematics, or computer sciences is required, though not necessarily in Hawking's subject. "It's not a research job," says one of his previous GAs. "I don't do any research of Stephen's, but I do help out with the public material and slides for technical talks, which is Stephen's research; it does help having a degree in maths."[68] Chris, another GA, confirms: "This is more an engineering job. . . . I do all the technical aspects associated with Stephen. [. . .] That's a soldering iron. . . . This is very practical thing. The sort of thing you learn in craft classes in schools rather than math classes."[69] In this job, "there is a lot of equipment, a lot of computing, a lot of media," which means engineering, logistics, and trip organizing. This is why a new candidate is generally interviewed successively by the PA (the memory of Hawking's extended body); the department computer officer; the current GA; and Stephen Hawking himself.[70] Similar questions will be asked concerning his experience with computers and his ability "to pick things up quickly"; his experience with travel and organizing large events;[71] his experience with disabled people (Hawking used to have physiotherapy before going home; as Chris said, "I'm picking up Professor Hawking! God, I don't want to drop him");[72] and his capacity to deal with stress, long hours, and intense work with other people. At the end, Chris thinks that he was chosen because he had some experience with disabled children, and the previous GA, Tom, thinks that it was because "he is relaxed with Hawking."[73] And he added, "Hawking said that he was quite confident, he needed someone like that when we're traveling because you do need to throw your weight around a bit." His assistant becomes his spokesperson. The job is extremely stressful (especially during travel), though sometimes interesting, even glamorous, given the celebrities one is likely to meet. As Chris says,

> The whole point about this job is that one minute I'll be doing one thing, and then I'll be doing something else. And then I've got to get back to the other thing. You know, you've just got to be able to change tracks seam-

> lessly, as it were, and be able to pick up—if you're doing A, you go over and do B, finish B, and go back to A like you never stopped doing A. That's why, you know, you don't do things in blocks, you have to swap and change horses. . . . Well, this morning, I was sorting out some travel insurance for Stephen's trip. His wife gave me a call, and said the screen has, on his laptop, it's gone a bit funny; we really have to sort something out. So I actually had to put the phone down and sprint to meet Stephen halfway, just sort of, because it's quite a serious problem actually, bring him back here, mend the computer, put it back in his chair, and then get back and get on with the travel insurance again. And . . . the point was, I had to have a reasonable idea of what I was doing before, so that, but that's just a small thing, I mean that's not a particularly grand example. Um, and, stamina as well; I mean, I've done a thirty-six-hour day.[74]

This is the reason the GA typically never stays for more than a year. And he adds, underlining the routinized aspects of the work, also "because it's not going anywhere, because it would be the same." The assistants must resist becoming automatons.[75] Moreover, Hawking's GAs tend to have ambitions of their own. Before they leave, they will be in charge of advertising their own job and of training their successors.

There is a relatively well-defined division of labor between the graduate assistant and the nurses. The GA assists with nursing sometimes; however, the nurses take care of nearly all of Hawking's physical needs, from brushing his teeth and feeding him to combing his hair and dressing him. And, if the GA can assist sometimes with nursing, the nurses can turn the computer off and on again and do very basic things like changing the power sockets over, because this is very simple. But they can't do something more complicated. During the String Conference, Chris said, "You just plan that [the computer] doesn't do anything wrong, but I always got back and I was never more than five minutes away."[76] Indeed, though he is occasionally able to take time off, the assistant is always on call. Though nurses are clearly the most important elements of Hawking's entourage, in this particular context, they rely entirely upon the GA to facilitate the continuity of their work.

The tasks are relatively well defined between the graduate assistant and the personal assistant as well. She is in charge of knowing if Hawking will go or not to the conference, when he will go, what he will do, and who he will meet, but when it is a question of booking the flights (he'll need to take three or four nurses with him) and the daily itinerary, "then the graduate assistant does all that, because he is going to be the one who is actually there at the time."[77] The nature of the job depends to a certain extent on who is in the post, as different people have different expertise. Some like to answer the phone, others don't; some are better at improving the computer system but less good at the paperwork; but in general the roles are relatively well defined. As Chris says,

> Once [Hawking] says: "Yes, [meaning] I'm going to do it," it's my problem; I have to deal with it. *It's an executive job.* If he says yes to something, it becomes my realm; before, it is not really my problem. If people want to come and ask Stephen to do something, they ask the PA—she's dean of the outside world. . . . I tend to make forays . . . out, every now and again; if somebody rings up, and I answer the phone, and they sound interesting, then I will do the whole thing from start to finish. But normally I won't pick up unless Stephen has agreed to do it.[78]

Reminding us that this network of humans and machines has become a well-orchestrated organization, Chris says, "*It will be a shock to the system* when she [the PA] goes. . . . But basically, it doesn't really make that much difference to me who does that job; as long as somebody does it. . . . From a logistics point of view, it doesn't matter. From a personal point of view, I wouldn't like her to stop because she is very good at the job, and her leaving is going to create a lot of work."[79]

Part of the GA's task is to help prepare diagrams for lectures: "For Stephen's lectures, . . . I sit behind Stephen. [. . .] The other lecturers just change transparencies, and Stephen obviously can't do that. And perhaps the advantage of being Stephen Hawking is that somebody else will do the slides for you, and that sort of thing, . . . sitting behind him. I mean, we haven't finished then yet; it's all . . . got some question marks."[80] But again, Hawking can't really explain. So the GA will draw something, and Hawking will say no, and he will draw something else, and Hawking will signal his approval. Chris gave me a detailed description of this process:

> It's really frustrating, actually, for the pair of us when you do it, because Stephen says, "I want you to draw a cylinder, and I want you to draw arrows between the top and bottom," but he can't really explain because it takes a very, very long time. So I will draw something, and he will say, "No, I want this," and it takes an incredibly long time to draw a picture, especially if it's a very technical picture; there was a graph on there that started, went up, came back down again, curved, and then got trapped between two lines, and he just said to Harvey [his student], "Draw an effective potential graph," and Harvey thought he wanted a graph to look something like that, but Stephen wanted one that did something else entirely, and he was very frustrated because, basically, he didn't want to have to . . .

> the reason he asked his student to do it is because he didn't want to have to go to the effort of actually explaining what he wanted, but he had to do [that] in the end, and that upset him a little bit. It wasn't my fault. He wasn't upset with me, but he was a little annoyed at Harvey, because he is his student; he is supposed to be able to draw a graph, but . . . I think it was actually because they were arguing [about] what Stephen should say in his lecture, I think, *just as physicists do*. Harvey drew the actual correct effective potential, what Harvey *thought Stephen was talking about*, [but] Stephen was talking about something else.[81]

The GA also replies to the numerous e-mails that Professor Hawking receives daily: "There's two e-mail accounts around Stephen. There's the public account, and there's the private account. That's the public account . . . *That's me again*."[82] Those concerning general scientific questions will be answered by the graduate assistant, if he has time and if he can, or they will be redirected to Science Net, "whose basic business is to answer scientific queries for the general public."[83] E-mails concerning Hawking's opinion on a scientific question related to his field of research are referred to Los Alamos (now Cornell), where one can visit the site on which all Hawking's papers are listed: "So you're in HEP-TH [High Energy Physics—Theory Archives], and you say ALL YEARS, and you just type . . . HAWKING, you do a search, and you will get all of Stephen's scientific papers that he has posted since 1991."[84] E-mails concerning a very specific field of Hawking's works (and the GA decides which are important) are redirected to his students, who help to clarify points or correct misunderstandings: "If it's really specific," says Chris, "and it's something I know—it's a sensible question, and somebody is trying to just clarify a very small thing—then I will redirect them to one of Stephen's students, who will write them back, and just say, 'No, you've got it all, you know, like . . . the idea of energy for black holes or for radiation actually works like this, not like that, which is a common misconception . . ."[85] Everyone gets a response of some sort, or almost: "You get somebody writing up, saying, 'I think Stephen is a complete w—,' just, you know, someone being very, very abusive, and you don't reply to that."[86]

Hawking replies to e-mails only very rarely, in exceptional cases, even though they are sent to an address bearing his name.[87] For example, one day his assistant received a desperate e-mail: someone wanted to commit suicide because she had Lou Gehrig's disease. The e-mail was redirected to Hawking's personal address, known only to his close colleagues, family, or friends. He immediately replied, "He was mortified that somebody was going to kill herself . . . also mortified that they wanted [his] advice."[88] "Obviously, people expect an answer from Hawking himself." As Chris recalls, "I got a letter back from somebody, saying, 'I wanted Stephen's opinion, not the opinion of some low-level bureaucrat. I don't care what you think.' I just ignored him. I was fed up. . . . I mean, the fact of the matter is, if you send a message to Bill Clinton, do you actually think he reads his e-mail? If you send it to clinton@whitehouse.gov, do you honestly think that Bill Clinton sits and reads his e-mail? No!"[89] Similarly, Stephen does have a telephone inside his computer. "[It's] a new thing [and] doesn't work quite well yet. . . . He can talk on that desk phone with his desk computer. It's all connected up so he can use it. But . . . a managing director with a phone doesn't answer his own phone. . . . No, when the phone rings, [the PA] or I would normally answer it. If need be, one of the nurses will. Stephen would not answer the phone."[90]

While the public wants Hawking's opinion in person, his close colleagues tend to perceive the assistant as being an extension of Hawking, in the same way that one addresses and looks at the translator instead of the person who speaks a foreign language: "Yeah, well, all the professors *know* that, that they're supposed to send messages to Stephen's personal account, but they seem to forget and send them to me anyway. So I just bounce them across."[91]

Finally, his GA polishes up his Web page: "Anything that goes DAMTP, cam.ac.uk, user Hawking, *is me*."[92] Of course, this "me" changes every year, but the address stays the same. And the Web pages haven't changed for a while, or at least the framework of the site doesn't change. The GA upgrades various bits, for example, by adding the title of the latest conference to which Hawking has been invited, a photo of Stephen Hawking and Bill Clinton taken during his stay at the White House, and the drawing of Hawking when he appeared in the *Simpsons*. But again, who decides? "Um, yeah, well basically, *it's up to Stephen what goes on the site. Sort of . . . Well, no, I don't generally discuss that much of what I put on the Web site with Stephen*, unless, unless there's something drastic."[93] Drastic! Like changing the page called a "Brief History of Mine" that Hawking wanted to put up. "That's been up there for about five years, and that's not going to change. If I was going to change *that*, I'd ask Stephen."[94]

In addition, the GA is supposed to explore new assistive technologies as they become available. He is also responsible for taking care of the equipment—for example, installing Hawking's software voice after it was renovated by Intel. Why a software voice? Because, as Chris says, "Everybody else, in the last three or four years, has been having software voices. I mean, I've got software voices. . . . They have this, why shouldn't *we* have one? And the difficulty has been getting one that sounds like him. Because you can have a software voice, it's easy, just give him a voice. But it doesn't sound like

Stephen."[95] When explaining that Hawking's old voice would be displayed in the Science Museum in London, he played what his new voice would sound like. Hawking's voice filled the room:

> *Hamlet said, "I could be bounded in a nutshell, and count myself the king of infinite space." I think what he meant was that although we humans are very limited physically, particularly in my own case, our minds are free to explore the whole universe. And to boldly go where even* Star Trek *fears to tread. But is the universe actually infinite, or just very large? And does it ever lessen, or [is it] just long-lived? How could our finite minds comprehend an infinite universe? Isn't it pretentious of us even to make the attempt?*[96]

THE TRIP

Stephen Hawking's identity as STEPHEN HAWKING depends on the proper functioning (the coordination and union) of this extended body. This is also true when he moves from one point to the other, to give conferences at Oxford in front of the general public, at the White House in front of Bill Clinton and his guests, or at Potsdam in front of physicists. This network of competencies allows a yes or a no to be operational. A yes to the question, "Do you want to go to this conference?" enables him to travel from one end of the earth to the other without having done anything more than twitching an eyebrow. The secretary and the assistant mobilize the entire context. They prepare the show in advance (the scientific part included). Part of this extended body, his secretary, stays in Cambridge, while the rest (his GA, three or four nurses, sometime his wife, the wheelchair, the computers, the batteries, the synthesizer) follow him from the beginning to the end of his performance. The principal role of his assistant is to ensure the maintenance of his flesh-and-blood body together with a variety of technical devices and to anticipate the unknown. Hawking goes up in the plane with a hoist, while the GA disassembles the wheelchair. The batteries and the computer follow first Hawking, then the wheelchair, or they stay and wait for another plane. The principal role of the GA is to maintain this part of the collective body—Hawking/machine/voice—together. For example, if, as has happened in the past, the wheelchair is not there when they arrive, things can become very complicated. Such problems tell us that Hawking and his technical devices are now one integral person.

> They didn't want batteries to go on [the plane]," explains his assistant, "and we ended up arguing on the flight itself; . . . they were saying it shouldn't go on, . . . [and we said] they obviously go on, they've got to go on, *it's for our*

> *flight*. . . . Stephen's got to have the batteries, it's not in dispute, . . . *he's got to be able to talk*, and now we get into a situation where they're taking it off me, and they're bloody minded, and I know it's allowed, and they're saying batteries are not allowed, I was showing the batteries, and they were taking them off me, and I just exploded on them, and Stephen thought it was quite funny.[97]

Jane, his ex-wife, recalls that

> even with the help of the nurses, assistants, and students, every such expedition was now indescribably nerve-wracking, attended by some twenty to thirty pieces of luggage and so many unforeseen factors that an expedition to the Himalayas would have seemed like a children's picnic in comparison. . . . While we held up a long queue, the check-in clerk would find it difficult to suppress his or her disbelief and irritated frustration at the amount of luggage *and apparently impossible demands of the frail but commanding body in the wheelchair*. Whatever seating had already been allocated, Stephen's requirements admitted no opposition, and accordingly hasty rearrangements would have to be made to seat him and his entourage wherever he chose. He [had] scant regard for the airline's normal policy for dealing with disabled people because he had devised his own. Come what may, he would insist on driving his wheelchair to the door of the plane, suspecting that to do otherwise might mean the loss *en route* of a vital part. Only there would he allow it to be dismantled, with strict instructions as to the disposition of the various parts. . . . He himself would then have to be carried on board. As often as not, the seat he had chosen would not be suitable after all, and he would want to change yet again. The laptop computer would then have to be set up so that he could argue out the details of his gluten-free meal—if indeed one had been provided—with the steward. The flight itself would usually pass with little disruption, apart from umpteen requests for warm water to wash down his medications—until, that is, Stephen bought himself an altimeter to check the cabin pressure. Woe betide the airline that tried to economize on fuel by minimizing its cabin pressure. Often the duration of the flight would pass in a flurry of messages from Stephen to the captain, conveyed by one or another of us to the cabin crew, insisting that the cabin pressure be brought to the equivalent of atmospheric pressure at six thousand feet. Each of these demands, taken singly, was reasonable enough, even justifiable; all together they would fray the nerves of even the most rugged traveling companion.[98]

The batteries, the computer, and the voice have become extensions of Hawking's body, without which he is unable to perform.[99] At the conference, the assistant is in charge of Hawking's meals, of driving him around, of taking care of the people who are going to welcome him, and the press. He will protect his private life, be in charge of the equipment, and, if Hawking has a spasm, he will immediately remedy the problem. According to Tom, "One of the main things is you don't get much time for yourself, because you are on call every night, and if something goes wrong you're on."[100] But as soon as the performance starts, the human parts of the extended body disappear . . . to reappear only at relevant times and places.

Blurring Boundaries: The "New" and the "Known"

OXFORD, FEBRUARY 24, 1998: HAWKING "SPEAKS" ON THE THEORY OF EVERYTHING

Eight hundred people are now waiting in the Oxford Town Hall. Tonight, thanks to the L'Chaim Society, Professor Stephen Hawking, "The greatest scientist since Einstein, is going to speak on the 'Theory of Everything.'" The crowd is full to overflowing. It is eight o'clock. The curtain is still down and won't be going up. Tonight everything will be played in front of it.

The stage is empty except for Thomas, the GA, who stands behind a laptop, plugging in an ensemble of electronic devices. Rabbi Botteach, friend and advisor of Michael Jackson, joins him. The lights fade. Silence. I hear a voice announce the entrance of "the biggest genius of all time: Professor Stephen Hawking." The wheelchair climbs a ramp made especially for him. Stephen Hawking sits in his wheelchair facing the crowd under the projectors. Behind him stands a white board upon which pictures will be projected. On his right, Tom waits. The rabbi leaves the stage. The conference starts. For the next two hours, pictures will follow one after another, echoed by a voice coming from nowhere. Everything is silent. Alone, this body upon which all eyes are fixed "speaks" about the theory of the unification of the universe. The lights come up, and it is time for questions. They have not been prepared in advance, but the show must go on. When the Professor takes his time to respond, Rabbi Botteach asks his assistant to explain how Hawking's computer works. The extended body becomes visible again. The nurse comes and goes to move Hawking's legs. Thomas, carefully and methodically, explains how Hawking uses his computer and his voice. On the white screen, the enlarged projection of Hawking's computer has replaced the talk about the origin of the universe; the public follows the cursor, which quickly

Figure 2. Hawking speaking at the White House Millennium Evening Lecture. Courtesy of the William J. Clinton Presidential Library.

chooses the words. Hawking is still immobile and silent. Hawking takes his time to respond. The public becomes agitated. The voice coming from nowhere will respond to three or four questions, for the most part about God. Rabbi Botteach announces the end of the show. Followed by his nurse, Hawking leaves the stage to a storm of applause. Thomas finishes packing the technical equipment and then follows. The next conference will be held at the White House. Professor Stephen Hawking will be speaking in front of Bill Clinton and his eminent guests for the "Millennium Evening." His talk, entitled "Imagination and Change: Science in the Next Millennium," will address how science and technology will shape and be shaped by human knowledge. The conference will be broadcast live on cable, satellite, the BBC, and the Internet. This time the extended body will stay invisible. It won't be necessary to fill the gap between the public's questions and Hawking's responses. This time everything has been meticulously prepared in advance. Hawking and HAWKING will be perfectly synchronized (see fig. 2).[101]

POTSDAM STRING CONFERENCE, JULY 1999: "THE UNIVERSE IN A NUTSHELL"

We can see this same process of synchronization as it plays out a year later in Potsdam at the International Conference on String Theory. Hawking begins his talk with his oft-repeated formula: "*I shall give this lecture with a computer and a speech synthesizer. I have come to identify with this voice, even though it has a weird accent*" (laughter). The next words he speaks seem more than a little bit familiar: "*Hamlet said, 'I could be bounded in a nutshell and count myself the king of infinite space.' I think what he meant was* . . ." Indeed, these were the same words I heard a month earlier when I was alone with the GA in Hawking's office in Cambridge. It is now reprinted in his book *The Universe in a Nutshell*.

Three days later, since the questions are neither "improvised" nor submitted in advance in Cambridge, the graduate assistant gathers—and prescreens—questions submitted by journalists before the press conference. The fifteen questions are narrowed down to three. Interestingly, the questions are always more or less the same. As the GA says, "If they actually did any research, they could just find the answer to their questions themselves—they don't need to ask Stephen. . . . [They] would ask *us* a question that Stephen's, that only Stephen can answer, rather than a question that anybody in this whole building can answer, including me."[102]

This time the GA doesn't offer a choice of menu to Hawking but is himself presented with one. He then takes the questions from his pocket and reads them to me: "In the beginning of the 80s, you expressed your hope that in the age of the new millennium, something like a 'theory of everything' [TOE] would have been established. Despite considerable progress, you don't understand much more than the basics. Also, string theory is still completely at the starts." The GA stops and comments, "This is obviously, written in German and translated into English, rather badly," and then continues with the question: "What is your prognosis concerning the TOE, [. . .] what does it look like today, now that we are at the edge of the new millennium?" "Well, basically," the assistant explains, "what they are saying is: 'Do you believe that we're going to find a TOE?' And Stephen has given a public lecture, a number of times, it's his standard public lecture, called 'The Millennium Lecture,' which he gave to Bill Clinton, where he is saying in twenty-five years. What Stephen will probably do to answer that question is just take the last paragraph of his public lecture, and say it."[103] He adds, "Because it's daft—it's a question that, again, it's a waste of time, it's not a waste of the journalist's time, because it answers the journalist's question, but in effect, it's a waste of Stephen's time. In fact, it would be much better if they just researched."[104]

He now reads the second question: "'The fact that particle theory and gravity theory have not been united until today might be unsatisfactory for physicists . . .'" "Now," the GA comments, "particle theory and gravity theory is just the theory of everything, it's the same damn question." He continues: "'In practice, however, this [is] a rather small problem, because both theories do not have points of contact, apart from black holes. Provided that there will be a TOE one day [Chris says, "Same question"], what would change (I guess in our common lives), except for a few textbooks?'"[105]

Chris says, "So question one and question two are basically the same question. And these are, in fact, questions that I got [the physicist X] to filter, so even he, to an extent, has put the same question twice. Um, and these, again, 'The search for the theory of everything has caught the attention [of] theoretical physicists all over the world. On the other hand, there are physicists who declare that such . . . [Chris: "Hang on, what are they saying? . . . Charlentry. It's not actually a word, but there you go."] for instance, Sir John Maddocks . . .'"[106]

"So," Chris concludes, "this question is, 'Do you believe we're ever going to find a TOE?' This question is, 'What do you think the effect of TOE will be?' and this question is, 'Do you think it's worth looking for a TOE?' [OK (laughs)] All questions can be answered in one. . . . He'll probably take the last paragraph of his lecture, and go 'dadadadadada.' And the journalists will be fans—they'll go, 'Wow.'" We will, moreover, see not only how answers are recycled by Hawking, but also by his assistants, who archive his responses so that they can use them to respond to other questions.[107]

Hawking "Responds"

A crowd of journalists is anxiously waiting. Hawking arrives in his wheelchair. Three physicists come and sit beside him. Silence. The press conference begins.[108]

> Host: Thank you very much. Now, we go to the questions, and we have prepared some questions that have been submitted, and Stephen has prepared an answer. And I think he should be able to answer those questions first, and I will read them. [And, confirming what the GA just mentioned,] I think these questions will be very similar to questions that might be asked anyway. So, I will read the first question to Stephen. "In the beginning of the eighties, you stressed your hope that on the edge of the new millennium, something like a theory of everything should have been established. Despite considerable progress, we still don't understand much more than just the basics. Also string theory is still completely at the start. What is your prognosis concerning the TOE? What does it look like, today, now that we are on the edge of the new millennium?"
>
> SH: *In 1980, I said I thought there was a 50/50 chance we would find a complete unified theory by the end of the century. Although we have made great progress*

in the last twenty years, we don't yet seem much nearer to our goal. So the lesson is that there is still a 50/50 chance that we will find a complete unified theory in the next twenty years, but that twenty years starts now.[109]

Host: Okay. The next question was, "The fact that particle theory and gravity theory have not been united until today might be unsatisfactory for physicists. In practice, however, this advice has a rather small problem, because both theories do not have points of contact, apart from black holes. Providing there will be a theory of everything one day, what will change? (I guess in our common lives), except for a few textbooks?"

SH: *We already know the laws that govern matter in all but the most extreme conditions. But we don't know why the laws are the way they are, or how they fit together. And we don't know the laws that hold in the extreme conditions at the beginning, so we don't understand the origin of the universe, or why we are here. This is what the human race has been striving to find throughout history. A complete unified theory might not bring much material benefits, but it will answer an age-old question.*[110]

Host: Okay, thank you very much. Now the final question that Stephen has prepared is the following: "The search for the theory of everything has caught a lot of attention among theoretical physicists all over the world. On the other hand, there are physicists who declare this search as charlatanry. For instance, Sir John Maddox. What do you answer to those critics?"

SH: *Theoretical physicists have been accused of playing mathematical games that had nothing to do with physics, because they can't be tested by experiment. But that isn't true. We have already observed quantum gravitational effects in the fluctuations in the microwave background, and we hope to observe supersymmetric partners, when the LHC* [large hadron collider—the CERN accelerator] *is working. Furthermore, if the ideal submerged extra dimensions are correct, we could have a whole range of experimental tests that could be carried out at accessible energies. Quantum gravity is real physics because it will be testable by experiment.*[111]

Host: Thank you very much. So I think that we can now open the questions. The first question is from this gentleman, please. These questions will be addressed to the panel of scientists accompanying Hawking.

It should perhaps come as no surprise that on the day after Hawking has performed and replied in person to these "three" questions, media from all over the world (BBC News, *Florida Today*, the *Deseret News*, *Princeton in the News*, the *Aiken Standard*, etc.) will recycle "his answers," over and over and over again.[112] Indeed, the authority of the most oft-repeated statement must be localized in this singular individual. In this sense, we have followed a socio-cognitive system, a network composed of humans and nonhumans, an adaptable machine, whose goal is to localize an utterance so that its authority is established. One simultaneously constructs the performance and the statement: "Hawking says." We see the inverse movement described by Latour and Woolgar in *Laboratory Life*, where a statement gains its authority as a fact only insofar as all its modalities disappear.[113] Moreover, though collective labor produces the statement, after the statement is recognized and becomes a fact, there is a split, and we decide that the collective is the consequence of the statement rather than vice versa. This presents us with a strange paradox: we repeat and recycle only because we valorize the unique, the original, and the authentic.[114]

Deseret News, July 21, 1999: Hawking "Said"

> HAWKING AWAITS UNIFIED THEORY PROOF
> POTSDAM, Germany (AP)—Stephen Hawking remains confident that physicists will prove string theory—a so-called "theory of everything" to explain the universe—but said it might take longer than he had expected.
>
> The world's best-known physicist, who was attending a conference on string theory, revised his prediction in the 1980s that there was a 50–50 chance the theory would be proven in 20 years.
>
> "Although we have made great progress in the last 20 years, we don't seem much nearer to our goal," he said. Hawking's odds on proving the theory are the same, but he now says it could take another 20 years.[115]

CONCLUSION

Sociologists, historians, and anthropologists of science have shown the complexity involved in the replication of experiments. In this chapter I have focused on the replication of STEPHEN HAWKING from the laboratory to the public stage. We have followed the slow and intricate translation of a twitch of an eyebrow into STEPHEN HAWKING "speaking" to the public. I have described how Hawking is "preprogrammed"; that is, how contexts of reception are selected, mobilized, and translated by his secretary and his graduate assistant; and, conversely, how, once Hawking has prepared his talk and his responses, the triptych—Hawking/computer/synthesizer—is transported from the laboratory to the public stage.

More precisely, I have shown how the collective surrounding Hawking mechanizes him or, put somewhat differently, transforms him into a kind of automaton that can only respond with a yes or a no. On the other hand, we have seen how this collective is itself a kind of machine that is organized in such a manner that a yes or a no can produce a HAWKING who "thinks," "moves," and "performs." His computer screen presents a menu and a choice; he clicks with his finger or not; his assistants (his secretary, his nurses, his GA) present him with a menu and a choice; he twitches his eyebrow or not. At a general level his yes's or no's make him move from one point on the planet to another. At a micro level, they make the machines and humans around him act.

Exegesis: Writing Hawking

I have shown the constant work of translation and exegesis carried out by this network. The machines, but also the assistants read and interpret "his" movements. They complete "his" words. They finish "his" sentences. They fill "his" silences. Sometimes they get it right. Sometimes they get it wrong. But in either case they do create meanings, discourses, and (inter)actions that allow Hawking, in turn, to generate texts, "give talks," and "respond" to questions. These too will be recycled—by Hawking and by the journalists who ask him questions, which seem to be echoes of prior responses, which are themselves (re)produced by (more or less the same) questions asked by other journalists in different (but more or less the same) venues. If Hawking is deprived of his computer, he is constrained to respond with a yes or a no by a choice on a menu. Paradoxically, the "freedom" that offers him the possibility of writing/speaking is limited by the repetition and recycling of the already said—both with regard to the computer program that allows and delimits his ability to write/speak and the journalists who induce him to repeat the same stories, anecdotes, and responses to their questions, again and again. Simultaneously, this repetition and standardization of accounts participates in the construction of his identity as STEPHEN HAWKING. We note here the optical coherence made possible by the writing process. This is the same *dispositif* that allows Hawking, the GA, the journalists, and the social scientist to compare questions and responses, and to match one to the other.

Hawking's Extended Body: Who's in Charge of What?

Contrary to what the public thinks—that is, that Hawking is pure mind—the role of his flesh-and-blood body is central to the activation of this network. A twitch of an eyebrow or a smile makes the collective function, a collective that is itself an extension of his body.[116] Indeed, everyone and everything in this collective (the machines, the voice, the assistants) is an extension of his body, and, simultaneously, each separately is in charge of one aspect of the construction of his being and his identity. For example, one is responsible for the media, another for the public performance, another for the machines, or the body, or the intellect, and so on. Indeed, the computers and programs extend the intellectual capabilities of Hawking and allow him to perform by endowing him with the capacity to write articles, go to conferences, and respond to questions (whether already prepared or not). The voice offers a double inscription as an instrument of thought. Indeed, what is well conceived is well said; this is all the more the case for Hawking, who doesn't speak. At the same time, this voice effaces—makes us forget—the role of the machine insofar as it speaks for, comes from, and marks the presence of a public persona, despite the fact that every utterance is written in advance (either by Hawking or by someone else). The secretary responds to the media, classifies and responds to the mail, and prepares his conferences. The GA is responsible for the mechanical parts of the collective: the computer, the wheelchair, the synthesizer. He is also responsible for the media: he responds to e-mails and phone calls, he updates the Web site, and when it is a question of giving a conference, he books the plane tickets, prepares the images and diagrams, ensures the transport and coordination of Hawking/machine/voice from one point to another while also taking care of the accommodations, food, journalists, and so on. He is in charge of synchronizing Hawking and HAWKING. The nurses take care of all of Hawking's physical needs.

Hawking's Competencies: The Collective That Acts

Hawking's competencies are in part exteriorized, distributed, and materialized in other bodies around him. For example, without his computer, Hawking can't write or communicate with someone who doesn't know him. But this computer is made for someone *like* Hawking who can activate its potentialities.[117] He can't make the mouse move, so the screen moves for him; he can't write with his hands, so the computer completes his words and sentences—it "guesses" what word he's going to use next.[118] He can't talk, so the computer or his assistants give him a voice. His assistants are chosen according to criteria of someone *like* Hawking (both academic and disabled). Then the assistants and computer attune themselves to the pattern of this user. They are all working toward the goal of reproducing the unique Hawking—fine-tuning, adjusting, and modifying their practices at each intersection of the network. Most of the time the assistants (and the machines) do their work without asking Hawking anything. The secretary decides what will be interesting for him (for example, she shows him invitations but not letters from his admirers). The GA decides much of what will go on the Web site; sorts and redistributes his e-mails (where Hawking's competencies are already inscribed, Los Alamos, for example) or sends them to his students, who in turn respond by translating "Hawking's" theories; or organizes the questions of journalists and decides which ones are the same or not, and so on. In other words, "they" are the ones who anticipate, project, complete, interpret, decide, classify, and act. However, only Hawking appears to be at the origin of all this work.

The Hawking Machine: Replicating Hawking

We have followed the functioning and arrangement of a collective composed of men, women, and machines whose tight dependency and coordination of action allows, at the end, the production of an actor who moves, writes, speaks, and performs. Hawking's capacity to act depends on the meticulous coordination of this collective, and so also does his identity as STEPHEN HAWKING. This collective body, his extended body, has become a system, an adaptable machine (the Hawking machine), a company (Hawking, Incorporated) whose goal is to manage, maintain, and preserve (keep track of) the authenticity of the speech of his genius. It repeats "what has never been said before": the unique Hawking is replicated. We discover a laboratory that produces immutable mobiles participating in the construction of an identity.[119]

As a vast, repeating machine, Hawking's extended body could be perceived as a form of externalized consciousness that is mechanical and automatic or, after Freud and Lacan, homeostatic. It seems impossible for Hawking to forget himself and to function as a subject by intermittence—that is, beyond performativity—as we all do sometimes. He is the omnipotent object of attention, precisely because he is so dependent. Hawking is an externalized deployment of ego that is not situated in face-to-face interaction. From this point of view, we must take into account the prostheses that lead to his formation. We'll see that Hawking seems more "three-dimensional"—more "real"—when everybody works around him, than when we are face-to-face with him alone (see chapter 5).

Hawking as Exemplum

Despite all this, one could say, to a certain extent, that there is nothing really specific about Hawking. Rather, his disability makes apparent the normally hidden practices of academics who shift the administrative aspects of their positions onto others and also the practices of managers who spend most of their time writing rather than talking, who lead their groups with a simple yes or no, who delegate competencies to their subordinates (the assistants answer the phone, book the plane tickets, and respond to e-mails). Also, his disability makes apparent the practices of movie stars who delegate competencies to people who market their image. Their position authorizes the possibility of these practices, when, conversely, these practices make possible and allow the maintenance and enactment of their position. However, in his case, his competencies are *more* distributed, collectivized, and materialized than those of anyone else. For Hawking, language—but also *his* intellectual competencies, *his* identity, and even *his* own body—become the property of a human/machine-based network rather than of the individual himself. His body is taken charge of, certainly, but it is also like the body of a king who is bathed and dressed, or a manager whose secretary makes him or her coffee, or a surgeon whose shoulder is scratched or whose forehead is wiped. If one can see in this *prise en charge du corps* a form of exemplification of other practices/identities (the king, the manager, the surgeon), one can see with Hawking a further supplementary degree of delegation.[120]

More Distributed than Others

Hawking's reliance on the textuality of the computer as his medium for talking and writing involves the blurring of the boundaries between the two and, accordingly, between the ephemera of lived experience and the archive. The frontiers separating the inside and outside of the mind are also blurred, since the locus of thought is made visible by the computer and the assistants, while written discourse comes first and is immediately translated into speech (we start with the screen, and we end with the perpetuity of writing); or again the boundary between the authenticity of a discourse—its novelty, its originality (the genius)—and its repetition: the already seen and the already said (a repetition made possible precisely because of our valorization of the original and the unique). Finally, the boundaries between the private and the public persona are also blurred (e.g., his secretary takes charge of his private life or his finances; his assistants become suddenly visible during a conference; and his computer appears on the screen shown to an audience instead of images meant to accompany his talk).

This collective of human and nonhuman actors works upon and selects the information that Hawking deals with. Accordingly, what arrives before his eyes is already totally prescreened. Conversely, thanks to this collective, he extends his presence under different forms (the accounts, the voice, etc.). One can, thus, follow the oscillation between the collective and the singular.

Where Is Hawking?

So, where is Hawking? Sometimes one witnesses the superimposition of HAWKING over Hawking—for example, when the GA says, "The public e-mail account: '*it's me*,' the Web '*it's me*,'" or "Why should *we* not have a voice?" or again when Hawking's colleagues start treating his assistant as an extension of Hawking and send the e-mails they intend for Hawking to the assistant instead.[121] One also witnesses moments where Hawking and HAWKING are not synchronized, as when people write to HAWKING and receive responses from his GA. In this case, they feel insulted; they want a response from Hawking (although if the GA had signed for Hawking, which he often does, they would have been fine). Or when Hawking takes the time to respond to questions that have not been prepared in advance and leaves it to his extended body to fill the space/time created by his silence so that the show can continue. To put it another way, what are the different manifestations of his presence? Is he a text stored in a computer? Is he a voice speaking in his assistant's office? Or an actor playing on stage? Is he a list of articles inscribed on the Web site of Los Alamos—"you type Hawking, and his articles appear"? Is he a door opening in his nurse's room?

Manifestations of Singularity

Where, then, does his singularity appear? Perhaps, one could say that Hawking's competencies and his assistants' competencies are not totally "interchangeable." This network is designed to produce an individual identity, but it is both when Hawking participates in this collective construction of himself and when he resists it that his agency appears. If Hawking takes the place of the user anticipated by Woltosz, he also appropriates these systems: he refuses to change programs, and thus his old programs must be supplemented and expanded (though he is now their only user); he refuses to alter his voice, though better ones now exist.[122] He has the last word when it is time to decide who will be his assistants. He decides where he wants to sit in the plane. He plays with his silence and with the audiences' expectations while on stage. He plays with the journalists when he explains he is having an identity crisis and expresses his desire to change his American accent (though he has refused all opportunities to do so). He doesn't want to change his biography on his Web site. He plays with his disability in the discourse enunciated by his voice that performs in his GA's office. He chooses a different response from the one his GA had anticipated.[123] He chooses what he wants people to remember about him, though we will see (especially in chapter 4) how difficult it is to resist the collective phenomenon of repetition.

Indeed, Hawking is far from predictable. As his secretary told me after I asked her if she knew how Hawking decides if he will give an interview or not, "I haven't the faintest idea. Sometimes I think he's going to say no, and he says yes, and sometimes I'm sure he's going to say yes, and he says no."[124] He plays also with the public during his conferences: "Sometimes, just for fun, he has been known to deliberately wait five minutes before responding with a monosyllabic reply. Audiences love it and burst into spontaneous laughter. On more than one occasion he has been known to wait five minutes, only to ask for the questioner to repeat the question."[125]

His assistant's reaction shows to what extent it is difficult to read him:

> I try to interact with him in the same way I interact with anybody else. It's obviously difficult. I think I can count on the fingers of one hand the number of times I've actually had a full-length conversation with Stephen. But you don't . . . he can't really have a conversation with . . . anybody, because it's very, very difficult. But I interact with Stephen very well, I think. I like to think we've got very friendly. I think he rather likes me, in fact. And, you know, you never really know, he could just hate everything I stand for.[126]

To conclude, one can talk about a kind of distributed intentionality, as when the assistant begins to identify with parts of the network: "the web, *it's me*"; "Why should *we* not have a voice?" In this sense, "to be," "to act," and "to do" become properties of a collective. However, these processes of distribution also create differences and participate in the construction of an identity: that is, it is through them that HAWKING and Hawking's singularity comes into view. We will follow this thread of thought into the next chapter and beyond.

II The Students

"How, with his severe disability, has Stephen Hawking been able to out-think and out-intuit his leading colleague-competitors, people like Roger Penrose, Werner Israel, and (as we shall see) Yakof Borisovitch Zeldovitch?" reflects physicist Kip Thorne. "They had the use of their hands; they could draw pictures and perform many-page-long calculations on paper—calculations in which one records many complex intermediate results along the way, and then goes back, picks them up one by one, and combines them to get a final result; calculations that I cannot conceive of anyone doing in his head. By the early 1970s, Hawking's hands were largely paralyzed; he could neither draw pictures nor write down equations. *His research had to be done entirely in his head*."[1] In this short passage, Kip Thorne, like a good anthropologist, sociologist, or historian of science, describes the practice of physics. Contrary to what we believe, he tells us that scientists think not only with their heads, but also with their hands. That is, they draw diagrams and manipulate inscriptions that can be reproduced at any time or at any place, recombined or superimposed indefinitely, all for the purpose of verifying long and laborious calculations.[2] Perhaps this is why Brian Rotman has proposed that we analyze mathematics as a practice—as the business of manipulating written signs. As he says, "Those things that are 'described'—thoughts, signifieds, notions—and the means by which they are described—scribbles—are mutually constitutive: each causes the presence of the other, so that mathematicians at the same time think their scribbles and scribble their thoughts."[3] But what happens when a theorist, like Hawking, does not have the use of his hands, does not—or cannot—draw pictures and perform many-page-long calculations on paper?[4] What can we say about theoretical work when no visible traces can be seen? Thorne, who seems genuinely puzzled by the invisibility of Hawking's work, concludes that he must do it all in his head. This is also the view of the general public:

> Physicist Stephen Hawking is confined to a wheelchair, a virtual prisoner in his own body, but his intellect carries him to the far reaches of the universe. —"Roaming the Cosmos" *Time*, February 8, 1988

And, indeed, Hawking himself comes to the same conclusion: "It would be difficult for someone that is disabled to be an observational astronomer. But it would be easy for them to be an astrophysicist, *because that is all in the mind. No physical ability is required*."[5] In the same vein, John Gribbin and Michael White emphasize that when Hawking discovered he had ALS, "[he] just happened to be studying theoretical physics, one of the very few jobs *for which his mind was the only real tool he needed*. If he had been an experimental physicist, his career would have been over."[6]

Theories in this view are produced theoretically, that is, by the mind alone—they have nothing whatsoever to do with practices, colleagues, bodies, places, and instruments.[7] In this respect Hawking, more than anyone else, seems to embrace the conception of the knowing subject as formulated by the rationalist tradition from Plato to Kant.[8] But how is this possible? How can Stephen Hawking generate new theories *on the strength of his reasoning* alone? Is it true that he needs nothing more than a "good" head to think? Is he a special—unique—case that challenges or sets limits to the results obtained (or the hypotheses made) by anthropologists, sociologists, and historians of science who put their hearts and souls into the effort to return scientific knowledge to the *collective* and *concrete manipulation of techniques* and *material inscriptions*?[9] I attempt to answer this question here by reconstructing the network of competencies that enables Hawking to do physics.

"BECAUSE HE DOESN'T DO EVERYTHING WE DO, HE HAS MORE TIME TO THINK"

Things that are normally taken for granted and seem entirely natural for those who have mobile bodies—like getting up, washing, dressing, and eating—are extremely time-consuming for Hawking. Everyday tasks would be impossible without the mobilization of his family and his nurses, on whom he depends entirely. Conversely, at least according to some, Hawking's disability saves him an enormous amount of time precisely because he's "freed" from mundane familial and professional duties such as taking care of the children or teaching and sitting on committees. Indeed, for many of his students and colleagues, this is why he can be so productive. As one astrophysicist remarked, "The pressures of doing anything, like blowing your nose, are very great if you have that degree of disability, and therefore, you know, it takes longer for somebody else to dress you in the morning... it takes longer to eat. I mean, it's a very difficult life, if one's disabled to that degree.... On the other hand, as I say, the contact hours with students and this sort of thing *probably* takes less time than it does for someone else, so there are compensations."[10] Or, as one of Hawking's colleagues said,

> The goal of our subject is that we are seeking one equation, which will describe everything. In a sense it's a very simple goal.... And so I think, in a way, you could say his existence, because he's, it's a very simple one, right? He doesn't do what other people do. He doesn't look after children. He doesn't cook dinner. I'm late today because I went to check on my daughter at school. He doesn't have that. He sits in his chair, and he has his computer. He can't talk much; that means he doesn't teach. He doesn't deal with the bureaucracy that university—all of us—have to. He has twenty-four hours a day to contemplate. And if you are looking for a simple explanation for something, that's a good way to be. You don't want too much noise.[11]

And finally, according to one of his students, "I also think that during the weekend (in part) he *probably* reads a lot. Like, I go and play tennis, or I go and cycle, or I go and take care of my kids, okay, he can't do that. He reads. So ... he has somebody to turn the pages, but ..."[12] But does the fact that he cannot carry out all these activities imply that he is a "pure mind," detached from all worldly and material conditions, that only "thinks," "contemplates," or "reads"?

In the first chapter we saw the fundamental role of a part of his extended body: a collective composed of humans and nonhumans, which works and does "in his place" what he cannot do for himself; thus he is enabled to be, to act, and to perform. We can also easily imagine that he does not teach or sit on committees; others do so for him. This is perhaps true for other professors as well. But does this necessarily mean that he has more time to "think," or that all the human and nonhuman "prostheses" that serve the indispensable task of maintaining his body and building his identity are absent or unnecessary when it comes to the mind or so-called "intellectual" work? In other words, does he need anything other than a "well-made" head to think? Once again, what is at issue here is the process of materialization and collectivization of thought at work in this *extreme* case. How does *he work*? Where is his laboratory? In his head? In his computer? Inside the walls of his office? Is there a collective, and if so, what precisely is it? To answer these questions, we first need to understand what *to work* or *to do* means (in theory) and what *he* relates or refers to. We will thus have to study the hands and the fingers, the eyes, and the context of those who produce knowledge. How does Hawking communicate with those around him (e.g., students, colleagues) and how do they communicate with him? What type of information is transmitted via body language and the machine? What kind of devices, protocols, and statements are produced? Could it be that, far from being detached from all collective and material conditions, he is more embodied that any other physicist?

By leaving the collective/extended body (nurses/personal assistant/graduate assistant/computer) that we explored in the previous chapter and shifting focus to another collective or another part of this extended body (computer/students/colleagues), I continue to examine (1) the possible exteriorization of Hawking's competencies, asking how far one can go in describing these processes of distribution; (2) the singularity or noninterchangeability of competencies (to see if there is something specific in his way of being or thinking, or if this specificity is a characteristic of the network);[13] and (3) the question of origins (do ideas come from him, or is he accumulating more and more selective and refined data that he then retranslates?). I also, moreover, begin addressing some additional questions. When we refer to the exteriorization of competencies, are we simply referring to how (socio-)cognitive tasks are delegated to others or made easier, or are we speaking about his intellectual competencies per se (that is, how they do the work)? Is it that his disability differentiates him (or empirically produces differences), or, on the contrary, does it exemplify how theoreticians work (whether his students or his colleagues)? How does his way of working, for example, differ from that of Albert Robillard, who has the same disease and is a professor of sociology and anthropology in Hawaii,[14] or Professor Reginald Golledge, a blind geographer?[15] Finally, I also continue to explore the question of a possible definition of individuality/humanity, of the collective body as a "metaphor" or not, and the differences between theory and practice, genius and assistants, humans and nonhumans.

THE PROGRESSION OF HIS DISEASE REVEALS THE PRACTICES OF PHYSICISTS

Before Hawking was surrounded twenty-four hours a day by nurses, a secretary, and a graduate assistant, his students acted as his nurses and assistants: "Because, I mean, being a student of Stephen involved much more than being a student of 'normal' people. I mean obviously . . . Stephen needed more help. [. . .] Nowadays he has 'round-the-clock nursing, but in those days he didn't, so a student would always help out a lot."[16] Living under the same roof, they took turns doing a variety of tasks. As Gribbin commented, "To earn their keep, they were expected to play as required the role of nanny, secretary, and handyman, helping with travel arrangements, babysitting the children, drawing up lecture schedules, and managing household repairs."[17] As one of his students told me jokingly, "Halfway through my PhD, I was wondering, like any PhD student, if I was going to graduate or not and get my degree. And I remember one day I was pretty discouraged, and I asked Stephen, 'If I don't get my PhD, can you write me a letter so I will get a certificate from the Red Cross, that I can do some-[?]' [laugh] At which, he grinned a little bit, you know? [Ha-ha-ha]."[18]

At the time, Stephen Hawking's wheelchair was not yet motorized, so his students had to push him around. Likewise, before he had a tracheotomy in 1985 and permanently lost his voice, Stephen had great difficulty expressing himself; only a few initiates could still understand and interpret what he was saying. His students served as his voice at conferences and also in the laboratory when he was dictating scientific articles. For Hawking to read, his secretary or one of the students used to hold the book or article in front of him. Sometimes the extended body expressed its suffering. As his secretary noted, "When they were all hard copies, then for a long time I used to hold them while he read them. Then *my arms got very stiff and my back got very stiff*. So then he had a music stand, and I just used to put them on the music stand and somebody would turn the page over when he finished."[19] To facilitate the selection of relevant information, his students pinned a list of articles to the wall that Hawking could consult at leisure or showed him each article in a pile of preprints as it arrived from Harvard or Princeton.[20] He would sort through them quickly by means of a nod (or a no): "We would just kind of show it, and he would say, 'No, no, no, no.' And then he said: 'Yes, ok,' and you would put this one on the side [laughter]. And then, 'No, no, no, no.'"[21]

What Do the Machines Do?

Today, thanks to his computer, Hawking can do all this by himself. Indeed, he can bring up a manuscript on his screen, read it, print it, or keep it in his private electronic library as easily and as efficiently as any other physicist. Inaccessible articles can also be scanned and put on his computer. Today, it has become a standard practice to send any new article (before or concurrently with its submission to a journal) directly to an electronic archive.[22] Thus, anyone can check to see who has written what, scan the titles and abstracts, identify anything new, and access the selected articles with a click. In this context, the computer has replaced the students: "All physics papers are now on the Internet. It takes just three seconds to click on the page, and you just read it. So there's no drawback there—*we all use the Internet all the time, so in fact you don't usually need libraries any more*. All current papers are on the Internet, and most information is there."[23] As one of his students notes, "The exchange of ideas is (*a*) very fast and (*b*) computerized, which are both very useful for [Hawking]."[24] New articles may also be sent to him directly, by e-mail.[25] Thanks to his computer, Stephen Hawking, like any physicist, can mobilize data at a click. The simple switch in his hand enables him to obtain feedback and to identify problems, interesting questions, and promising methods; it also allows him to gather them together to find new answers.

Today Hawking's students like to point out that they would rather perceive or know him as a scientist, "which is entirely different from how the graduate assistant and his nurses know him."[26] In fact the students have become students again, while what I have called his extended body, the collective comprised of humans and nonhumans on whom he depends entirely, has multiplied by division: on the one hand, the personal assistant/graduate assistant/nurses and, on the other, the students/colleagues. If the students interact with Hawking's secretary or his graduate assistant (GA), it is from a purely organizational or administrative point of view, "but that doesn't really affect us too much," they say.[27] Every time one of them publishes a paper with Hawking, he sends it to his secretary so that she can archive it. Every time they need to know where Hawking is or will be in the next few months, they contact his GA: "So it's booking Stephen up for various things."[28]

Yet, even though Hawking has once again become a being endowed with "a voice" and a certain form of autonomy—owing to the use of a computer that enables him not only to surf the Web and to read and mobilize data at will, but also to write and communicate—he always works in very close cooperation with his students. As he likes to remind us, they are his closest collaborators. This is partially due to his disability, for Hawking relies on his students to perform his calculations (as I will show below). But his disability also makes visible the practice specific to a theoretical physicist.[29] Far from being a discipline in which one can devote oneself to the joys of solitary thought, theoretical physics requires intense collaborative work.[30] As one of Hawking's collaborators puts it, "Often people ask me what it's like to work with [Hawking], and I don't find it unusual at all, because *in theoretical physics one very often works with many different people, very different*. And I don't see Hawking as kind of that different. And you, I think people from other fields don't understand, I think, because *theoretical phys-*

ics is very collaborative. And so we're always, as it were, changing partners (laugh). And that's hard to understand."[31] One might conclude, then, that all scientists are in some sense "disabled" insofar as they are unable to think without being attached to a set of instruments, machines, and collaborators. In this sense, does it really make any difference if the body of the scientist can't move, or if he has "a joy stick and a talking computer on his wheelchair"?

Although it has become standard practice in the community of physicists to cooperate via e-mail, Hawking, surprisingly, prefers to work face-to-face (or side by side) with his students. Whether alone or with his machine, he is unable to do calculations or diagrams without them. Indeed, as we will see in chapter 3, it is because of Hawking's disability that we can appreciate the importance of the mobile body in mathematics and analytical reasoning. This reminds us how essential a mobile body is in writing mathematical symbols, working on a computer, or drawing diagrams.[32] The role of his students, therefore, is fundamental.

What Do the Students Do?

Hawking is usually assisted by four students whom he sees regularly:[33] "He's always willing to spend time with his students. So, let's say that out of the four students, there will be someone seeing him every day, at least one of us will be talking to him every day, or thereabouts."[34] His students are, respectively, in their first, second, third, and fourth years. Acquiring a PhD takes four years, and every time a student leaves the circle, he or she is replaced. Far from simply being Hawking's hands, legs, and arms (in a sense, the executive body of his mind), the students are all exceptional. We have here a form of impure Cartesianism.[35] At Cambridge there is a famous exam called "Part III: Certificate of Advanced Study in Mathematics," which draws students from all corners of the globe. One hundred are selected. After nine months of preparation, the best four will have the privilege of being interviewed by Hawking. Only one is chosen.[36]

EXTENDING HAWKING'S RESEARCH SUBJECT

Depending on his students' skills and "mutual" interest (from the Part III exam, he knows their strengths and weaknesses), Hawking assigns them different subjects so that there is a minimum of overlap between the subjects chosen. This might explain his capacity "to multiply his fields of investigation," to have an "overview" as his students say. In general a newly recruited student continues with the same research subject as his or her predecessor:

> I was fortunate, I think, that I approached Stephen and said that I was interested in *this* subject *that coincided with the fact* that another student that Stephen was working with at the time . . . was also working on this, and he was finishing. And I didn't know that at the time. *So I was just at the right place with the right problem at the right time.* So when I said that I was interested in this, *it just so happened that a vacancy was coming up.* So in that way, I did choose my project, but it's not always like that.[37]

Simon is studying the problem of virtual black holes and how they could affect predictability; Raphael is exploring quantum cosmology and black hole propagation, "one of Stephen's pet subjects," on which, they say, he has always worked: the origins of the universe and how it was created—the "Holy Grail."[38] "And I'm very happy actually in having been introduced to the latter subject. But, yes, he was looking for someone in that subject [quantum cosmology], and I was looking to do the subject, and *so it was a nice coincidence.*"[39] Bob is working on the mathematical formulation of general relativity, and Peter on "the stability of string vacuum." Although they share the same office and some complicity, there is little intellectual exchange between them; they interact directly with Hawking. Indeed, none of the students really seems to know what the others are working on. As one of them said to me, "There is surprisingly little interaction about *our physics lab work*, mainly because we are working on quite different things. [. . .] Stephen is a *sort of a hook*, and we all sort of go to him when there's a problem, but we're all working on quite different aspects of his ideas."[40] In this way Hawking multiplies the fields in which he is involved. This might explain "his" capacity to see the big picture, as well as his way of working by analogy.[41]

The practice of assigning research subjects to his students is, in itself, standard in physics. It is by working on specific problems that students train their minds and bodies; that is, that they learn the skills and tools necessary to become researchers. Thomas Kuhn's notion of normal science, and Michael Polanyi's concept of tacit knowledge allude to this.[42] However, students are trained in distinctive ways.[43] Unlike other PhD supervisors, who prefer simple exercises as a form of learning, Hawking has always tended to make his students work on extremely complicated problems: "Stephen drops his students in at the deep end from the outset. . . . He gave me a titanic task; that's why I never finished, and most of his students also have unbelievable tasks to perform, fundamental problems, which generally are accessible [only] to senior people, but [students] *have to adapt ideas and his way of thinking to these fundamental problems*.[44]

Even though the fields appear to be different, it seems that they are not so in the mind of Hawking, who endeavors to apply a certain working method to these subjects.[45] For instance, at the beginning of any collaboration, he asks his students to read the same book, *Euclidean Quantum Gravity*, a collection of articles he edited with Gary Gibbons.[46] This ostensibly provides the students with a common ground—that is, it puts them all on the same page—while at the same time it helps transform this text into part of the "literary canon" of physics.[47] Each student is then responsible for working on the problems or results left by his or her predecessor—problems related to furthering Hawking's "discoveries" or "ideas." From these they must try to draw conclusions, to "prove" or "demonstrate" their validity, and to apply them to other fields.

Tim, for instance, has continued the work of his predecessor, Simon, on virtual black holes, an extension of Hawking's work in the seventies on the quantum properties of black holes:

> Initially I worked on a project . . . investigating the idea of something called virtual black holes. The idea is that when you only have a quantum system, you get fluctuations at a quantum level. And for quantum gravity, that would involve black holes, which are very, very dense packets of energy, so dense that light can't escape from them; they are actually spontaneously popping in and out of existence on a very, very small scale. So *our research* is in centering on some ideas behind that, and also *what sorts of consequences that would have*. And one of the consequences which that has, and in fact, black hole evaporation full stop has, is that you have the possibility of losing information, which is a loss of quantum coherence. *And Stephen has been trying to demonstrate* that this would definitely happen in a full quantum theory of gravity for many years now. And a lot of people don't believe it, because it's actually quite a big statement to make. And so *he's trying to*, well, *rather we're trying to give a very persuasive argument for that case*.[48]

Thomas, by contrast, is pursuing Hawking's work on understanding the origins of the universe and, more specifically, on the consequences of the proposal on the initial state of the universe that Hawking formulated with Jim Hartle in the eighties.[49] "[Hawking] wrote some important papers on quantum cosmology in the early eighties, and *we* are trying *to ex-*

tend the ideas within these papers in the larger and the broader context of cosmology."[50] But this time the idea emerged at a conference in Chicago: "We [Hawking and I] went to Chicago. So then we heard, in Chicago, [that] Andre Linde was also there. He talked again about this eternal picture. And then, in one or another way, this idea came up: 'Maybe we should try to link, or try to place this eternal picture on firmer footing. Try to predict it from some fundamental quantum theory.' And that's how the second project started, and we came back from Chicago. *That's a short history of my past effort.*"[51] But only Hawking (the "hook") could make the connection. As Thomas says, "But I think no one is having the idea to link both things. And Stephen is an ideal person to have this idea, because he knows the quantum cosmology theory very well. And he knows about the eternal picture, so he says, 'Look, this eternal picture, it must have a translation within our quantum theory of cosmology.' So that was kind of the start of this project a couple of months ago."[52]

Professor Hawking appears to be the very incarnation of the Popperian vision of science.[53] He has, as one of his students notes, "the intuitive notion of what could be done," "the ideas," and "the inventive mind" that his assistants, tirelessly working on proof, try to put into mathematical form. The professor cannot do these calculations by himself for the simple reason, so often overlooked by epistemologists, that in order to do mathematics one needs not only a head but also ten fingers to write equations, draw diagrams, and use a computer.[54] Totally paralyzed in his wheelchair, Stephen Hawking can use only the single finger that his nurse delicately places every morning on the switch linking him to his computer. But a computer cannot calculate on its own; it needs to be operated by a researcher with nimble and dexterous movements. As Raphael, one of Hawking's students notes,

> *It is something that, no matter how you do it, it is very useful to have ten fingers that you can move, even if you do it on a computer*. And so it really would be inefficient for him to do that when he has students who can just, you know, you might have to explain to them what they're supposed to do, but *that is still faster than actually doing it himself. And, being in the position that he is in, he will always be able to get the people to do it for him.*[55]

Hawking also uses a program that transforms words into symbols, for although he cannot produce calculations, he can construct sentences, albeit succinctly.[56] As a result, his students do most of the work. But, again, this is standard practice in physics: PhD students or post-docs often do the bulk of the technical work.[57]

TRANSLATING HAWKING'S IDEAS

Design Phase

In the first phase of designing a project, the students attribute the initial idea to Hawking. He may give them two or three articles to read and sometimes the model to follow in performing their calculations, and then send them off to work alone. As one student said, "At the beginning of my PhD, he would give me a project. He would say, 'This other student has done this, and we should go and look at this . . . area. We need this, as a kind of start-up, and then, after that, you can start doing calculations in a direction by looking at this model or that model.'"[58] Another said, "He just basically told me the idea, and he told me a couple of papers I could look at, and references. But *mostly* he just has this intuitive notion of what could be done, and it is his students' job to put it into a mathematical description."[59]

Hawking also asks his students to prove that other physicists are wrong. This is how, some say, he motivates himself. He likes to debate important physicists, to argue and disagree with them, and to attempt to prove them wrong. Roger Penrose recalls,

> I can't quite remember exactly how this went. . . . It was a new student, just starting with Stephen. And Stephen was suggesting some problem for him to work on. And *this problem was to show that something that I had done was wrong. So the student had to do this*. . . . Then I went to talk to Stephen. And then I came out of the room, and I found the student waiting there, rather nervously. And he came up to me, and he asked me, "What is it I'm supposed to do?" Because he . . . it was quite funny.[60]

Sometimes an idea emerges during a conference. In this case, Hawking and his student will spend two months sitting next to each other "talking" about the physics behind this project and their intuition. "He tells me about his idea. And *I try to understand him, I add things*, and in our dialogue we are trying to formulate a project. So that's based, that's just *based on physical intuition and our thinking*."[61] The method of working is "back-and-forth" scientific discussion. But, when one has a scientific discussion with Stephen, one has to be patient, because it takes him a long time to make a sentence. One of his students adds, "He also tends to be very careful in choosing his words, and he chooses as few words as possible to describe something because it takes so long to say it."[62] Consequently, his sentences are enigmatic: "Whenever I talk to him, *the whole thing remains rather fake*. We never manage, or we rarely manage, to go *into detailed things*. So that *he's leaving that to his students, I think*. But, of course, he uses his computer to speak, so we can talk about whatever he wants. And so he can also describe in words what he means by his ideas."[63]

The comment that "he can't go into details " is made over and over again by his assistants, students, and journalists, and so on. This forces everybody around him to make things explicit.[64] In a sense, this goes against the notion of tacit knowledge and embodiment so dear to science and technology studies. However, one can also say that the theories they work on are not well defined; in this regard, Hawking can't go into details and can only propose vague explanations.[65] This is perhaps also why his students all agree that it takes time to adjust to working with him; one ends up getting used to him, but it takes time:

> TH: I guess when you get to know him better, it's nicer to work. Of course, it takes a bit more time than usual to get to know him.
>
> HM: What does it mean "to know him"?
>
> TH: Well, to feel a bit more at ease with him, I suppose. . . . What he expects from you, and . . . the way he likes to work with you, and all these things. It takes a while.[66]

When I asked another of his students if he had learned to read Stephen's body language, he replied,

> Oh, yes, definitely, you get used to that after a while. Stephen can raise his eyebrows when he means yes, and he can shake his head when he means no. At first you don't notice that, but very soon you get into that. I mean, people have been watching when I've had a conversation with him, and . . . all they've seen is me questioning and then going "Right. Okay." And how did you get an answer? But you can see . . . nods and shakes, so yes, there is definitely, there is a limited amount of body language there.[67]

The problem Thomas is currently working on is extremely difficult to formulate in a mathematical framework that might enable him to solve it, "but it's always hard to formulate these ideas in quantum cosmology. I mean *the theory is not so well defined* in the sense that it's hard to do calculations with it. So *he's, we're* trying to formulate these ideas in all these things."[68] At this stage of the process, when they are spending a lot of time thinking about the problem and making calculations, Thomas just goes to see Hawking in person if he needs him. He will communicate with him via e-mail only when they are really in the final stage of the project, namely, when *they* are writing the paper. "This project we are doing now has been very difficult to formulate in a mathematical context, in which we can actually start to solve, and almost the whole *process of formulating* really is due to Stephen. Neil and I, we were trying to follow him, basically."[69]

But once again Thomas cannot see the role that he is playing in Hawking's work. Because of Hawking's disability, his office becomes a center of collaboration:

> And I have the impression Stephen has kind of developed his mind to be able to work just without anything else, without equations, without someone else talking to him. But, on the other hand, I must say he takes his time. And when Turok and I, when we want to talk to him about this project, this can easily last for four hours in his office. And we will write down equations on the blackboard, and we will all talk about different aspects of the project, and we're trying to make progress. And ... he rarely says ... that we should go. And it can take [long], I assume that he must enjoy that as well; he must learn something himself.[70]

Thus, Hawking is often conceived of as an agent who performs operations that the others cannot, like "redirecting the conversation": "So, often, a very fruitful way to [work] is you have Stephen and, say, two other people, and there's a conversation between the two other people going on, trying to understand a problem, and also with Stephen, but Stephen's contributions to the discussion *are very slow, but usually very deep*. So you'll have a continual conversation going on, with Stephen once per minute, or once every two or three minutes, *making a statement that may completely change the direction in which the conversation is going*."[71] Or else his colleagues and students, like his assistants and his computer, offer him a choice of menus or ask him a series of questions to which he answers yes or no:

> But we all talk the same language, and ... communication is pretty easy in that context.... Social interactions are not crucial. It's good to have a sense of humor. It makes the work go easily, and Stephen has a very good sense of humor and makes jokes, and that makes it enjoyable. But ... I've never really had a problem communicating with him. I know we understand each other very well; there's eye contact. And he's able to say yes very easily and no very easily, and to me, that's enough. You know, because I can suggest an idea, and he will say, "No, that's wrong." And then I will think about it a bit more, and [say], "Is it wrong for this reason?" And he'll say, "Yes," or "Is it right?" or whatever.[72]

Problem-Solving Phase

During the problem-solving phase, the students are left to their own devices with the little training they have received. At this stage Thomas works with his second PhD supervisor, Neil Turok: "I think that you are supposed to be quite independent, as far as your calculations go, if you work with Hawking, you know, computer work or whatever. . . . He has been quite useful in just presenting new ideas and thinking about things."[73] Christophe, a recent student of Hawking's who is continuing Tim's work on the loss of information (although with a very different approach), prefers to roam:

> It's a bit my problem, if you like, because the purpose of a PhD supervisor is to orient his student towards something that works. *If you like, Stephen's orientation is very slight, although it's very deep*, but it's not. . . . It's rather me who'll look in the direction of the things I think of myself, and since I'm a novice in the field, I bang into walls most of the time. [. . .] The . . . first thing that he asked me to do concerned correlation functions in black hole decay to zero, black hole space-time, that's all. I didn't know what a black hole space-time was. I'd had some lessons at university, vague notions, but nothing really practical, and correlation functions, I didn't really have a clue as to how to find that. And *so all these words, I had to find out one by one* what they all meant from a slightly deeper perspective than the partial way in which I'd learnt before. And I encountered an impossibility I couldn't manage. I was unable to do the calculation.[74]

Only after they have tried everything do the students go to Hawking. He is a last resort. He sometimes solves their problem very simply. Christophe explains:

> So well, yes, I told him that I didn't understand and so on, and instead of carrying on, he changed my subject, *and he asked me to calculate something else that was happening there in infinity in another theory*, and in fact that was even more difficult. It's impossible to calculate that; nobody has ever managed, ever. So we reverted to the idea of the first calculation, and there I ended up finding a calculation that I could do, that I could understand, and that I haven't done yet. But I've got it tucked away, and I should be able to do it by the end of the summer. In the meantime I did . . . I moved on to other things, if you like. Either something he asked me to look at or something interesting I found along the way.[75]

This is probably what Hawking means when he says, "I always work on several problems at the same time; that way when I get stuck on one problem, I can make progress on another one."[76] This is the same reason his students work on completely different subjects.

For the students, managing alone—either because the subjects are too difficult or because Hawking is unavailable—often means starting to talk to other professors (usually the most eminent, as the subjects are so difficult), initiating cooperation with them or following a different track that includes their own interests and questions that emerged earlier.[77] That is how they "progress." Every month the students present their work to Hawking, continually feeding him with new ideas and research questions.

Finally, in the last phase, when the students have results, they show them to Hawking. That is, they literally demonstrate them before his eyes. A simple yes or no from Hawking could result in months of work on their part:

> I would go into the calculation, and I would come back and say, "That's what, the calculation I've done, that's what I get." And, if he agreed, then I would pass to the next step. [. . .] If he didn't agree, then I would go to the blackboard and do it. And I would do it step-by-step, so he *could see* the step-by-step, and if he doesn't agree on one step, then I would say, "Why? What's the problem?" And if I'd made a mistake, then we'd say, OK, I made a mistake there, and that's how it would work.[78]

The Writing Phase

During the writing phase, the students are again left entirely to themselves. Although they do most of the work, they attribute paternity of the ideas to Hawking. This was the case with Raphael, who published two articles with Hawking; he wrote the first paper himself, whereas in the second one he partially reused elements from one of Hawking's talks. After this, the students discuss it with him, but there are generally no major changes. "The first one I really wrote on my own, but the idea was all Stephen's. The second one I wrote most of, but I also used some parts of a talk that Stephen once wrote, and gave, and I sort of recycled some element of that talk for the paper. And then when it's done, we'll discuss whether he thinks it's good or not, and if he'll want to make any changes, of course. But usually there won't be anything major."[79] Later Raphael added, "Well, what makes it a privilege to work with him is the fact that you can rely [on the fact] that what you will be working on is going to be original and very interesting. *It's unlikely that you'll just be recycling some other thing, and just doing some more*

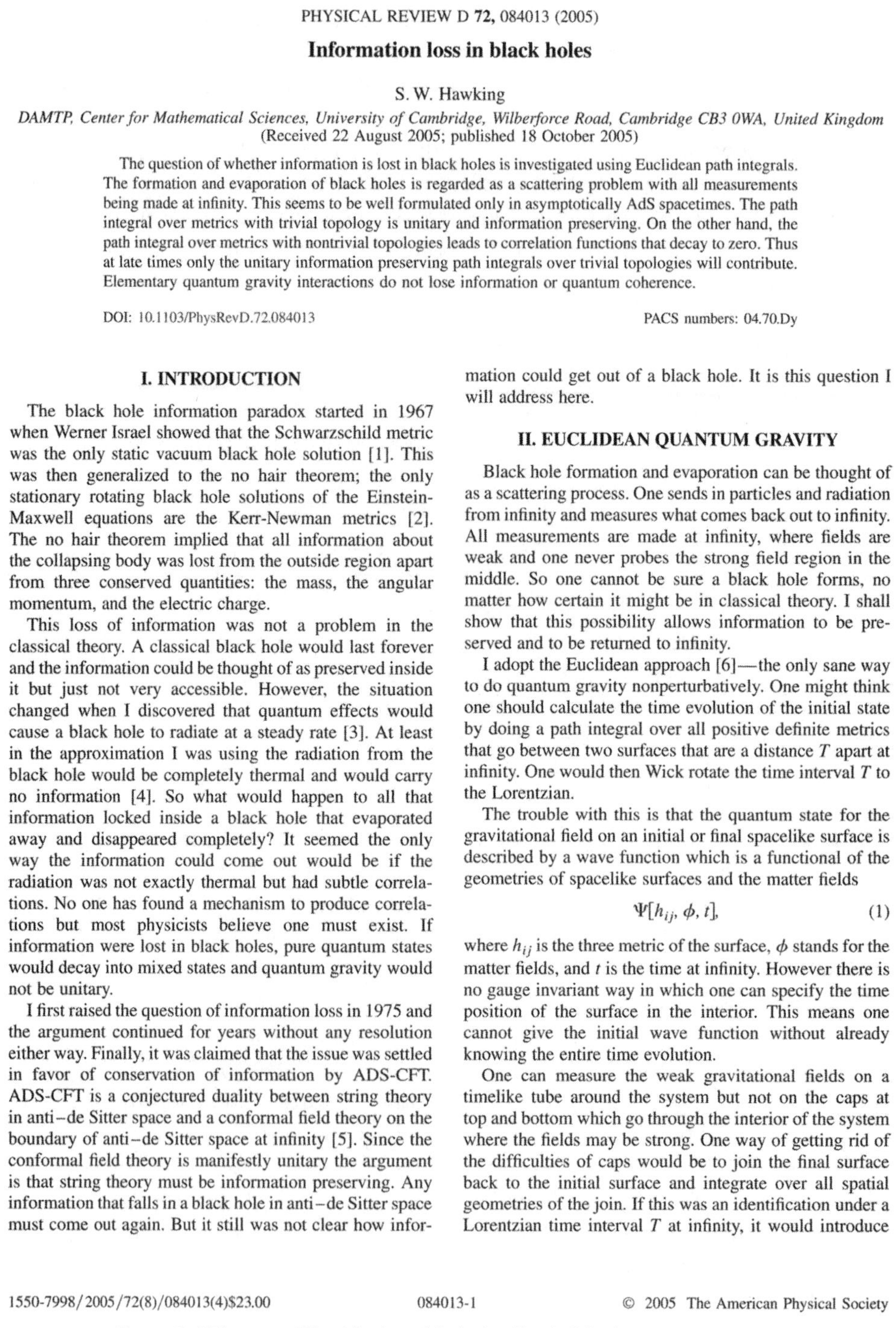

PHYSICAL REVIEW D **72,** 084013 (2005)

Information loss in black holes

S. W. Hawking
DAMTP, Center for Mathematical Sciences, University of Cambridge, Wilberforce Road, Cambridge CB3 0WA, United Kingdom
(Received 22 August 2005; published 18 October 2005)

The question of whether information is lost in black holes is investigated using Euclidean path integrals. The formation and evaporation of black holes is regarded as a scattering problem with all measurements being made at infinity. This seems to be well formulated only in asymptotically AdS spacetimes. The path integral over metrics with trivial topology is unitary and information preserving. On the other hand, the path integral over metrics with nontrivial topologies leads to correlation functions that decay to zero. Thus at late times only the unitary information preserving path integrals over trivial topologies will contribute. Elementary quantum gravity interactions do not lose information or quantum coherence.

DOI: 10.1103/PhysRevD.72.084013 PACS numbers: 04.70.Dy

I. INTRODUCTION

The black hole information paradox started in 1967 when Werner Israel showed that the Schwarzschild metric was the only static vacuum black hole solution [1]. This was then generalized to the no hair theorem; the only stationary rotating black hole solutions of the Einstein-Maxwell equations are the Kerr-Newman metrics [2]. The no hair theorem implied that all information about the collapsing body was lost from the outside region apart from three conserved quantities: the mass, the angular momentum, and the electric charge.

This loss of information was not a problem in the classical theory. A classical black hole would last forever and the information could be thought of as preserved inside it but just not very accessible. However, the situation changed when I discovered that quantum effects would cause a black hole to radiate at a steady rate [3]. At least in the approximation I was using the radiation from the black hole would be completely thermal and would carry no information [4]. So what would happen to all that information locked inside a black hole that evaporated away and disappeared completely? It seemed the only way the information could come out would be if the radiation was not exactly thermal but had subtle correlations. No one has found a mechanism to produce correlations but most physicists believe one must exist. If information were lost in black holes, pure quantum states would decay into mixed states and quantum gravity would not be unitary.

I first raised the question of information loss in 1975 and the argument continued for years without any resolution either way. Finally, it was claimed that the issue was settled in favor of conservation of information by ADS-CFT. ADS-CFT is a conjectured duality between string theory in anti–de Sitter space and a conformal field theory on the boundary of anti–de Sitter space at infinity [5]. Since the conformal field theory is manifestly unitary the argument is that string theory must be information preserving. Any information that falls in a black hole in anti–de Sitter space must come out again. But it still was not clear how information could get out of a black hole. It is this question I will address here.

II. EUCLIDEAN QUANTUM GRAVITY

Black hole formation and evaporation can be thought of as a scattering process. One sends in particles and radiation from infinity and measures what comes back out to infinity. All measurements are made at infinity, where fields are weak and one never probes the strong field region in the middle. So one cannot be sure a black hole forms, no matter how certain it might be in classical theory. I shall show that this possibility allows information to be preserved and to be returned to infinity.

I adopt the Euclidean approach [6]—the only sane way to do quantum gravity nonperturbatively. One might think one should calculate the time evolution of the initial state by doing a path integral over all positive definite metrics that go between two surfaces that are a distance T apart at infinity. One would then Wick rotate the time interval T to the Lorentzian.

The trouble with this is that the quantum state for the gravitational field on an initial or final spacelike surface is described by a wave function which is a functional of the geometries of spacelike surfaces and the matter fields

$$\Psi[h_{ij}, \phi, t], \tag{1}$$

where h_{ij} is the three metric of the surface, ϕ stands for the matter fields, and t is the time at infinity. However there is no gauge invariant way in which one can specify the time position of the surface in the interior. This means one cannot give the initial wave function without already knowing the entire time evolution.

One can measure the weak gravitational fields on a timelike tube around the system but not on the caps at top and bottom which go through the interior of the system where the fields may be strong. One way of getting rid of the difficulties of caps would be to join the final surface back to the initial surface and integrate over all spatial geometries of the join. If this was an identification under a Lorentzian time interval T at infinity, it would introduce

Figure 3. Title page of Hawking's article in the *Physical Review*.

calculations or . . ."[80] Thomas started communicating with him via e-mail in the writing phase only. In the end, he will write the paper and then send it to him. Hawking will read it on his computer; perhaps he will have comments or ask for revisions.

Christophe hasn't published with Hawking yet, but did co-author the paper he presented in Dublin with him. In such cases the work is done "with four hands." That particular time it was Hawking's student, not his computer, who completed the sentences. Hawking will start a sentence, and his student will finish it, or the student will propose a sentence, and Hawking will approve or not, with his face.

> HM: But this seems crazy to me; it seems very difficult to . . . how can you anticipate exactly what he's going to talk about; I would think . . .
>
> CG: Well, you know, the subject of the talk is, *it's what I do.*

Or else:

> HM: I wanted to read this article by Maldacena.
>
> CG: I wouldn't recommend it.
>
> HM: No! It's too difficult?
>
> CG: I don't recommend it at all. But you can read Stephen's; it's a bit clearer.
>
> HM: Is it?
>
> CG: (pause) . . . since I'm the one who wrote it (laughs).[81]

When the paper is finally published, only Hawking's name will appear on it; the only mention of the student will be his or her e-mail address at the bottom of the last page. We see here a double movement, for while his students attribute "the ideas" to Hawking, they also do all the work (see fig. 3).

CONCLUSION

In this chapter I reconstructed the network of competencies consisting of men, women, and machines that enable Stephen Hawking to do physics. We have seen that the scientist is by no means the disembodied mind that he is described to be. In fact, his disability highlights the incorporation required to perform theoretical work. *Because* it forces him to delegate his competencies to machines and individuals, his disability, far from making him an exception and a mindless body, reveals what is normally hidden or overlooked, like the role of instruments (here, the computer) and the students as loci of mobilization of information and collaboration. Moreover, what is usually attributed to his unique intellectual capacities is actually the result of a collective that has been handpicked and trained in his methods by the DAMTP. This collective informs, proves, calculates, recontextualizes, extends, and translates "his" discoveries. These discoveries are often articulated and elaborated with reference to work previously published by Hawking (which was also written with other students). Once written, "new" papers are in turn fed into his computer so that they can be used and (re)translated again. Yet there is nothing exceptional here; one can imagine other professors working in similar ways.

Indeed, this pure mind, far from being isolated, makes visible the material and collective practices of theoretical work. It reveals the hierarchical organization of students and supervisors in the laboratory, the way topics are assigned (ensuring the continuity of work done by previous students), and the mechanisms of authorship whereby supervisors are credited with the work students do (regardless of their actual contributions). Indeed, the supervisor acts as a kind of spiritual father or guide—a role linked to an intuition based on years of experience and developed through a sense of the "big picture." The supervisor picks the problems, decides what is interesting, and then leaves "the details" to the students, who will do the calculations if not write the whole paper. Often "the author" is little more than an editor.

By effecting a process of additional delegation, however, Hawking's disability illuminates not only that which is normally present but not seen, but also the fact that his competencies are even more collectivized and materialized than those of others. Indeed, in his case, all the motor and cognitive operations that normally go unseen by the ethnographer, either because they concern the private person of the scientist or because they are normally incorporated in a single body (as, for example, an able-bodied physicist), become visible. They are, in effect, externalized and incorporated in other bodies (technicians, assistants, machines, etc.). Accordingly, it is possible to "see" the work of delegation, multiplication, and redistribution of skills to people and machines so necessary for understanding the functioning of his mind (or that of any physicist).

Hence, our reconstruction of Hawking's laboratory, based on the relations and practices constituting it—who is who; who talks to whom; who does what, how, and why—has made an invisible part of his body visible. But can we stop there? Can we replace the savant's competencies with those of a collective that the ethnographer's reconstruction—the work of writing and assembling—has revealed? Does HAWKING-the-great-researcher resemble in this sense Pasteur as described by Latour? Is he the equivalent or simply the product of a collective that constitutes, so to speak, his body and mind?[82] While the computer makes it possible for Hawking to read and to mobilize information, and while the students enable the work of justification, calculation, proof, and writing, it is Hawking's yeses and nos (and a few sentences—via the computer) that redirect the conversation. They enable the students to move on, to see where they've gone wrong or how to proceed, to publish as is, or to amend. I have shown this process in which the actors constantly reveal what they are doing, that is, the collective at work: "It's our job to calculate," "I know what he's going to talk about *because it's what I do*," and so forth, and simultaneously attribute all this work to the ideas and singularity of a specific person: "He's got ideas," "He's got intuition"; that is, "He's got the capacity to identify interesting problems and to guess what the solution will look like (so that he knows where you have to look to solve these problems)." "He's got experience," "a really all-encompassing view"; "He thinks in terms of concepts rather than detailed calculations." This phenomenon can partially be explained by the work of projection, completion, exegesis, and elaboration that I examined in the previous chapter: the students are trained in such a way that they can interpret what he does, but in order to do this, they have to adjust; they finish his sentences, retranslate or formulate problems, are briefed by other professors, do calculations, and so forth. But, again, is this not what characterizes any manager or laboratory director at "the head" of a team, or any experienced scientist? As Woltosz recalls on the basis of his experience as a manager of a company,

> Well, I think managers are more generalists than specialists, even if they're managing a special department in a large corporation, they're still more of a generalist than the people that work under them. *And the higher you go, the more of a generalist you must be*, and *the more you must see the big picture and how everything balances with everything else*, or needs to, in running an organization. So I think, I read once, that when we're younger—you know, Einstein did his best work in his twenties, and things like that—when we were younger, we were very good, and I think of myself, back in those days, I was very, very adept at calculus and at very deep mathematics. Now I haven't done it in years, and *I tend to think more as a generalist and more in general concepts*. Now Hawking has made that evolution as well, from, you know, in his twenties when he still had his motor skills, solving equations and thinking in terms of very complicated mathematical relationships,

to a more conceptual thing. And I believe Einstein indicated that he went through the same sort of evolution. So there is really something to it.[83]

Let me recap. I started from the following processes of attribution: "He doesn't do everything that we do because *he can't* (e.g., administrative tasks and teaching), so he has more time to 'think,' and for that purpose he simply needs a good head." I have shown that even if he had more time to think, producing theories is the product not only of a good head but also of a well-organized laboratory or a department. In this sense Hawking's disability does not prevent him from working in his chosen field, for others will perform all the necessary operations. Once again, his disability acts, for the ethnographer, as a kind of magnifying glass that reveals what we don't normally see. It does not stop him from doing what he does, provided that others do what needs to be done. Some, including Hawking himself, even think this makes him particularly competent in his field. In the next chapter, I attempt to pursue the processes of attribution singularizing Hawking by taking my analysis a step further, asking whether my observations—which are partially the fruit of my own projections and partially a window onto similar practices (of academics, managers, and physicists)—tell us something specific about the functioning of this type of theoretical practice or about the ways in which certain theoreticians think (visual versus analytical), or about the way Hawking, in particular, does what he does. Once again, let's open another black box—what *does he do*? And *where* is he? Let's return to the laboratory.

III The Diagrams

SH: I regard my disability as a minor detail like [being] colorblind. If you are colorblind, you develop tricks to recognize traffic lights and clothes.

HM: What are the tricks for you?

SH: Pictures.

HM: In your mind?

SH: Yes.

Stephen Hawking, interview by the author, July 2007

INTUITION VERSUS CALCULUS

"Hawking is his mind." What characterizes him is his intuition, his capacity to identify interesting problems and to guess what their solutions will look like. This is what one hears in the corridors of the Department of Applied Mathematics and Theoretical Physics if one asks his colleagues and students about him. "Since he can't do calculations, he has to think. That's logical!" says one of them.[1] He seems to have the ability to see "through calculations" and to take "shortcuts." As Jim Hartle puts it, "Stephen is exceptionally clear, and he ... more so than almost any other ... theoretical physicist, understands *in a sharp way* what the issues are and ... how to have them resolved. So he can sort of *see through calculations*, for example, when the rest of us are typically often confused, you know, going one way or the other. It's a great talent and *very useful in this type of physics, which is far from experiment*."[2]

While the intellectual competencies attributed to the scientist seem perfectly suited to the type of physics that he practices, Kip Thorne, like many others, also sees a direct correlation between the progression of Hawking's disease and his way of thinking. Inevitably, by depriving him of the use of his hands, his disease has also deprived him of the ability to perform the gestural manipulations required by calculative work. Paradoxically, his handicap makes visible the fundamental role of a mobile body in analytical mathematical work: "Without being able to *write* equations down, it's not really possible to do long complicated calculations with equations. He can do amazingly well with equations despite that, but *he can't do as well working with equations as most other people at the top of the same field*."[3]

People say, as does Hawking himself, that he has transformed his handicap into an advantage and developed a way of *short-circuiting* his limitations by redefining questions so that he can *visualize* either their solutions (so that he can *see* what they would *physically* look like) or the frameworks within which problems are posed. In this way, he can solve problems by mentally manipulating images (forms, geometric structures, sophisticated models) rather than equations. He has thus become particularly good at finding solutions to problems that he can retranslate geometrically and less good at those that defy this kind of transformation. As Kip Thorne told me,

> Well, the big difference between him and most people is that he thinks about things very geometrically. And, whereas most people work things out analytically, by manipulating equations, this means that certain kinds of problems, *those kinds of problems that can be formulated in a geometric way, he can solve much more quickly*, he can understand much more deeply and very quickly, than anybody else. Certain kinds of problems cannot be readily formulated that way, and those kinds of problems he is less likely to make major discoveries on, although occasionally he can, through this method of having students write equations on the blackboard, staring at them, thinking about them.[4]

Hawking seems to confirm this intuition and to justify his way of thinking—and the fact of being able to do physics—not by affirming, as he did earlier, that the body is unnecessary for doing theoretical work, but by eliminating in one fell swoop the supposed foundations of scientific method and rationality: mathematical language. No one needs equations to do physics, writes Hawking, for fundamental ideas can be explained in words and images:

> The physical laws that govern the universe are usually expressed in the form of mathematical equations. For most people, this has created a great barrier to understanding. But equations in physics are like the financial

> appendices to the budget: important if you are an accountant concerned with the details but unnecessary for a general understanding of what is going on. *The basic ideas in physics can be explained in words and pictures.* [. . .] It is difficult enough to imagine objects in the three dimensions of space and one dimension of time that we are used to, let alone the seven or more extra hidden dimensions that may be there, according to our unified theories of everything. Still, one can generally ignore most of these dimensions and *just picture things in the two or three dimensions that our brains are capable of visualizing*. So I believe it is possible for everyone to understand the basic laws and forces that govern and shape the universe.[5]

EYES, HANDS, AND DIAGRAMS

Up to this point, nothing seems to challenge the premise on which a very Western conception of theoretical activity is based: that theory is fundamentally different from practice.[6] Thus, contrary to an experimental type of science—especially physics—that is seen as the fruit of concrete *work* necessitating manual dexterity, tricks of the trade, instruments, places, and precise gestures,[7] theoretical activity is assumed to be the fruit of mental, contemplative, and introspective operations alone.[8] This distinction, of course, is artificial. Theorization is not the result of inaccessible mental processes, but—as in the case of experimental science—a deployment of concrete and incorporated abilities based on the manipulation of theoretical technologies or paper tools.[9] Indeed, I intend to show that what seems to be the product of a purely introspective mental visualization—the mental manipulation of images—is, at least for Hawking, partially based on the physical act of seeing two-dimensional representations. In other words, his ability to bypass the limitations imposed by his disability—by redefining questions in such a way that he can visualize either the solution or the framework in which the problem is posed—is largely based on the use of material, concrete and "visualisable" devices, that is, on diagrams. As one of Hawking's students commented, "By using these diagrams, one can remain within the diagram and think in terms of the diagram, and still have a valid scientific approach. In other words, the diagram is equivalent to a mathematical equation. In fact the diagram is mathematical. That's it. Had diagrams not been discovered, by Penrose or Feynman, Stephen's life would have been far more difficult."[10]

However, the fact that Hawking needs a visual medium to think highlights the particular nature of his *métier*. As pointed out by Brandon Carter, the inventor of the diagrams named after him,

> In my opinion . . . *[that's how] physicists' brains work. Perhaps other humans are different, but ordinary physicists and mathematicians, it's because of our eyes, we're used to seeing things in two dimensions.* [. . .] For example, if you tell me that a car's going to travel at 20 km per hour at first, then accelerate to 30 km per hour, then go further down the road and finally slow down and stop there, I have difficulty remembering all that. If you do a position diagram, . . . it's far easier for me to remember the image of the diagram containing the same information, because the diagram simply says that it goes faster or slower.

And this need for visualization is inherent in theoretical practice: "That's why it's so surprising for us," says Carter, "when we see literary scholars giving lectures without using the blackboard. We, we . . . can't discuss anything between ourselves without a board or a piece of paper, . . . we're completely dependent on it—to communicate with others, but also to communicate with ourselves, to think about our own ideas, we need a visual medium. . . . I remember much better visually, . . . but *Stephen is the extreme case*, he's absolutely, he's even more dependent on this."[11] Diagrams therefore offer Hawking possible ways of seeing through problems that mathematical formulas obstruct. But, once again, how can he proceed? Unlike his colleagues, not only has he lost the capacity to write equations by hand; he can't draw diagrams either. The role of the students is therefore fundamental, for it enables us to see the processes of delegation required to understand (his) theoretical and visual practice.

Hawking has claimed that he was lucky in his choice to become an astrophysicist, for, as he put it, "to be an astrophysicist . . . *no physical ability is required*."[12] This is far from being the case, however, for to think, one needs one's hands or, if this is not possible, the hands of others. But above all, one needs eyes—eyes to see these hands work, write, calculate, and draw. Just as he has developed ways of selecting information provided by the human/machine collective that surrounds him—by raising an eyebrow (or not)—he can also mobilize and select information from diagrams drawn before his eyes. His eyes thus become the point of articulation and activation of a network of competencies composed of individuals, machines, and theoretical tools.[13]

More often than not, he asks to see Penrose-Carter diagrams—a direct application of the global methods invented by Penrose in the 1960s.[14] As Hawking's student says, "*This technique of representing* an entire space-time in a little diagram like this, this is something that Stephen uses all the time. As soon as one speaks to him about a certain universe, a certain space, a certain solution to Einstein's equations, *he asks* ... for the Penrose diagram *every time, because* this, *this is visual*."[15]

I pause here to reflect upon the inscriptions that the scientist manipulates.[16] What are the properties of this diagram? What does it do? Or what does it make the scientist do? The Penrose-Carter diagram is a device that makes it possible to *see* "the entire universe" in a finite image—which, as we shall see, is itself the direct translation of an equation that can be fairly complicated. Armed with the microcosm of the diagram, the scientist can dominate the entire universe with his gaze:[17] "A lot of time we *are talking about* spaces that are infinite, just infinite time-space; many of our models are about infinite spaces. *It's hard to have infinite pictures* [laugh],[18] but *part of the cleverness* of the Penrose diagram is that ... it distorts the space so that ... *all infinite space is mapped onto a finite-sized picture. The finite-sized picture still shows all infinite space*."[19]

Moreover, the competencies of the diagram, "its cleverness," enable the scientist to know in the blink of an eye the causality of an entire universe, that is to say, to see "who's talking to whom, who's talked to what, and why."[20] "Also, it distorts the space in the way which makes light travel in straight lines, *which is very useful for understanding how different points are connected to each other*, because nothing can travel faster than light."[21] Thanks to this device, it is possible to take shortcuts: "It's extremely practical, and, for someone who also knows how to use diagrams, it is very powerful. *You don't need to do forty thousand calculations; you can see it*."[22] Thus, the diagram is not an answer in itself but a heuristic tool: "It's the visualization of a space ... that becomes intuitive [if seen] from a certain angle."[23]

In other words, the global point of view—the connections, shortcuts, and intuitive capacity that we attribute to the power of the mind—are, in part, the product of these tools.[24] Gilles Châtelet, for example, argues that diagrams work as prosthetic devices that become vehicles of intuition and thought.[25] He calls them "techniques of allusions."[26] Yet the intellectual properties of the scientist are not only due to these tools, for if Hawking's disability does not prevent him from thinking visually, some physicists who are more analytical than visual say that they themselves are "handicapped" or limited when they have to do work involving visualization. In this sense, Hawking's disability makes a "diagrammatic" style of thinking apparent. But, once again, as we have seen, while he cannot use his hands to write and manipulate equations, it is also impossible for him to draw these diagrams. His students must therefore do this for him. Thus, to work with him, not only do they need to develop a way of communicating with him by reading the language of his body; they must also learn to talk and to write in the language of diagrams: "Yes, it's clear that every time we [present him with] a solution, a certain space that *someone has found*, ... the first thing that he asks for is a Penrose diagram. So, of course, at first you don't really know what it looks like ... for such-and-such a solution to Einstein's equations, but if you know that you're going to be asked a specific question, you learn to answer it."[27]

The task is by no means simple. As David Kaiser has shown, "Feynman diagrams were not an automatic representational scheme. They were 'conventions' that had to be taught and practiced, sometimes more explicitly than others."[28] One must also learn how to read and draw Penrose-Carter diagrams. The ability *to read* a Penrose-Carter diagram requires no superhuman skills—in fact it has become such a standard tool in the study of gravitation that any postgraduate student learns to master it from a theoretical point of view.[29] The student needs to know why it is used and how to obtain it in an equation. However, it is quite another matter *to actually draw* it. The difficulty is related to the fact that one first has to know the mathematical equation (that is, the analytical solution) of a space before being able to transform it into a diagram. In other words, the diagram is a possible translation of this solution: "You have a space that you know by its equation, and it's thanks to this equation that you manage to construct the image of the thing, but doing this the other way around, this is far too complicated."[30] And, for every new space discovered, there is a new diagram that corresponds to it. As Hawking's student says, emphasizing both the plasticity of this tool and the need to learn (and create) by doing, "I know by heart perhaps about fifteen different spaces because I've seen them, because I've understood them and worked on them. But you give me a new space, ... there is a Penrose diagram for this also. And that's a thing that *you've got to work on and do*. It's not something you find in books. And *if it's you who discovered the space*, it's not likely to be in a book."[31] Or, as he said in an interview the previous year, "For example, I know that Stephen asked [one of his other students] to look for something ... in relation to a space, say a space-time, in which there are two black holes which accelerate in relation to each other, and this is super difficult to visualize, to see. [...] *This is a problem that's easy to define but that's very, very difficult to solve.*"[32]

In short, to be able to visualize, there first have to be calculations—calculations performed by an elite group of agile, hardworking, and highly trained students. It is *they* who "discover" these spaces and translate them into diagrams. Calculation and demonstration seem to precede the idea and the intuition, which is just the opposite of what Popper argued. In a way, it is as if the context of discovery—attributed to a single mind—could take place only *after* the context of justification.[33] Thus, we could say that

Hawking's competencies are more distributed than those of other people, as he cannot do calculations himself. Therefore, we say of him—as he says of himself—that he is more visual than analytical. But, once again, visualization is in turn based on the use of diagrams that he cannot draw himself. At the same time, the diagrams we are referring to here can only be drawn *after* having found the mathematical equations for which they are projections. The student is therefore responsible for doing *both* the calculations *and* the diagrams.

This is where the complexity of the process seems to reside. In this respect, it is interesting to note that these diagrams seem to be perfectly adapted to Hawking's disability.[34] Whereas the strength of Feynman's diagrams lies in the fact that it is possible to draw many of them very quickly so that there is a constant feedback loop between the diagrams and the algebra (a direct line-by-line translation),[35] Penrose's diagrams are the result of pages of calculations that, as we have seen, can be done by the students even if Hawking is unable to do them. A diagram can be equivalent to a hundred lines of algebra.[36] Brian Rotman argues that diagrams "distill action and experience and 'reveal themselves capable of appropriating and *conveying all this talking with the hands*' . . . of which physicists are so proud."[37] We see that there are indeed real differences in the ways that one can *physically* manipulate diagrams. More than just an aid or prosthesis used to facilitate calculations, as in the case of the Feynman diagram, the Penrose diagram entirely substitutes for the work of calculation—it sums up all the information needed in one picture: "The Penrose diagrams are more often ways to characterize a space-time rather than calculate a hundred iterations of an approximate expression. . . . The Penrose diagram is functioning as a *portrait*, almost like *a cartoon*."[38]

This is where the superiority of the diagram resides, compared to a long demonstration consisting of pages and pages of equations.[39] As Hawking's student said, referring to the problem of causality, "You don't need forty thousand calculations, you can see it."[40] Finally, every time he is shown a new diagram, Hawking can memorize it: "And once he's seen it once, he's got it in his head."[41] This is how some explain his productivity:

> If you can't write it, you become better at holding diagrams in your brain, and to be willing to . . . look . . . carefully at different parts of it. It may help a lot to put the diagram on paper [laugh]. . . . I'm sure it's hard. My guess is [that] what he does is a combination of having someone draw the diagram for him on a piece of paper or a blackboard. And other times, [he] would . . . try to keep the diagram in his head. *It's certainly easier to try to keep the diagram in your head than a complicated equation.*[42]

In this sense, Hawking's handicap highlights a form of reasoning peculiar to certain so-called visual theoreticians, which he seems to take to the extreme. As one physicist put this to me, "So perhaps people like Hawking and other good physicists, you know, they just take that further, and they somehow internalize these sorts of diagrams so that they become even more powerful."[43]

Hawking has perhaps pushed this work of memorization of diagrams further than others because he is unable to lighten his cognitive tasks by drawing them or choosing other methods. As Brandon Carter says, "He can *only do* what we'd prefer to do." However, the diagrams cannot just remain in his head. If this were the case, nothing would happen. They have to be mobilized, drawn again, rendered visible before the scientist's eyes every time a problem arises, either because he asks for it, or because his students have been "trained" by his requests to anticipate it. Indeed, we are far from the image of the solitary genius here.

The diagrams take shape and have meaning only when they are mobilized, that is, in the interaction between Hawking and his students. They correspond to a given problem, intervene at a particular moment, and concern specific individuals. Moreover, as Brian Rotman, referring to the work of Gilles Châtelet, says, "Diagrams, figures of contemplation, live in the aftermath of the body's mobility. Their genesis and hence their description, action, and significance are inseparable from gesture. . . ." Rotman speaks of the "propensity for a gesture to awaken other gestures."[44] This is also why a standard diagram displayed on Hawking's computer screen would be of no use to him. Thus, just as the choice of menus proposed by his computer or the questions posed by his assistants enable him to communicate with only a slight movement of his eyes, so too in drawing a diagram before him, his students lay out the foundations of discussion:

> Yes, but it's as if . . . you [were] at a painting class, and the first thing that the teacher says is to take out your paper. That's really it. You take out your sheet. You want to work on what? So okay, you take out a sheet. You want to work with which brush? Okay, you take out the brush, and the teacher will see straight away whether you have a brush, a pencil, a charcoal, or something else. *If you put that there, and do the groundwork, Stephen knows immediately what you're going to talk about. You lay the foundations of discussion*. That's it.[45]

Just as a painter chooses the materials on which he is going to paint, the scientist chooses the diagrams from which he will do his calculations, or at least his assistants make available for him the material necessary for the "*mise-en-calcul*": "It's as if you [said to] an artist, 'Draw me a picture,' the first thing he'll ask *you*, he'll ask *himself*, is 'What paper am I going to use?' . . . 'Choose a canvas, a cardboard, a slate,' that's it. And then *you will* do the calculations within that. Afterwards you'll paint over it. That's more or less it."[46] Hawking likes the Carter-Penrose diagram because it is a direct representation of the space in which he is going to work. He thus thinks with, through, and within it. We can talk of the construction of a virtual reality with which the physicist, or a version of himself, interacts or enters into contact. As Elinor Ochs, Sally Jacoby, and Patrick Gonzales put it,

> In the physics laboratory, members are trying to understand physical worlds that are not directly accessible by any of their perceptual abilities. To bridge this gap, it seems they take *embodied* interpretive journeys across and through see-able, touchable, two-dimensional artifacts that conventionally symbolize those [physical] words. While in some cases the members do not actually touch a representation, they may journey to some part of it by gesturing along a delineated trajectory or toward a particular point, even at some distance (e.g., while seated at a table). In this sense, [their] sensory-motor gesturing is a means not only of representing (possible) worlds but also of imagining or vicariously experiencing them.[47]

THROWING THINGS WITHIN THE UNIVERSE FROM INFINITY

The Carter-Penrose diagram enables Hawking to see *which space he is in* and to grasp the rules that define it, which in turn allow him to identify what can be done and to visualize the obstacles that will greet him. This time the physicist is not only compared to an artist who is offered a palette of theoretical tools, but to a race car driver visualizing a map that will enable him to anticipate his course. It is clearly a question of a journey and of corporeal movement. As Hawking's student comments,

> If you ask someone, a Formula 1 driver, before a race, the first thing he wants is the state of the track, [he wants to see] what the turns are and all that . . . on the track. *And well, . . . the Formula 1 driver will want to have a map of the course, and Stephen wants the Penrose diagram*, as that's what enables him to have ideas, to think, and in a wink of the eye and a glance, most of the time, not always, because sometimes they are more complicated than that, but more or less at a glance, you can think and *see what's happing inside* a space like that.[48]

The diagram thus makes it possible to do easy thought experiments. As Châtelet puts it: "The diagram never goes out of fashion: it is a project that aims to apply exclusively to what it sketches. This demand for autonomy makes it the natural accomplice of thought experiments."[49]

Like a novelist imagining his characters, the designer of Hawking's interface, Walt Woltosz, put himself in the future user's place. As he says,

> I spent thousands of hours in front of a computer with one switch, writing letters to people that never got mail, . . . just as practice, . . . carrying on mock conversations with an invisible partner; . . . someone walks into the room, and I imagine carrying on a conversation about some subject. So, I would start using his switch, and I would picture them interrupting [me] and so on. I want to say, "Please wait till I finish what I am trying to say." . . . I wanted to be able to say that right in the middle of the sentence that I was trying to build, and so I figured out a way for it to do that. What we call instant speech. All of these strategies basically came out of being—the user is what I call it.[50]

Just as Woltosz identifies with his user as a means to create his program, so too Hawking projects and displaces himself in the diagram. As one of his student-collaborators says, "The Penrose diagram makes it possible to do thought experiments easily where *you throw things within the universe from infinity*—you see what happens. You can do so in your mind, using the diagram."[51] Here the diagram is rendered dynamic by a linguistic construction that implies a movement of the body—"*You throw things* within the universe from infinity."[52] This grammatical construction situates the physicist in the liminal world between visual representation and the constructed worlds that it indexes.

The example that we have given seems to confirm what Ochs, Jacoby, and Gonzales have described so well: "In this world, scientists engaged in collaborative interpretive activity transport themselves by means of talk and gestures [or imagination] into constructed visual representations through which they journey with their words and their bodies." Above all, "in this liminal world, sharp boundaries are not drawn linguistically between subject (i.e., researcher) and object (i.e., the physical phenomenon under study).[53] Indeed,

> when members of the physics laboratory journey across visual displays through talk and gesture (or in their imagination), they may construct themselves grammatically and somatically both as subjects engaged in interpretative activity and as objects of interpretation. Utterances such as

> "When I go below in temperature" [or "you throw things from infinity in the universe"] deconstruct the distinct social identities of scientist and scientific object, constructing in their place a blended, indeterminate identity. Like the image of the cyborg, this conjoined social identity conflates animacy and inanimacy and thrives in a liminal zone between here-and-now interaction, visual representation, and represented physical worlds, rather than in any single constructed world.[54]

Indeed, we shouldn't forget that the famous black holes on which Hawking works are thought experiments: "There *seems* to be evidence that black holes *might* exist, but there is a difference between the astrophysical black holes that we *see* and the ones that we *model*, which are very, very perfect, with a great deal of symmetry involved."[55] Black holes may exist, we are told, but there seems to be a difference between those that one sees, those that one *believes* one sees with elaborate experimental tools, and those that one models, that is, that one visualizes and experiments with using complex theoretical tools. Physicists thus construct mathematical models of our universe by using—when they are more visual than analytical—certain diagrams as a crutch or canvas upon which they can visualize problems and do calculations. In a sense, one can say that they never leave the models, which make it possible to produce questions that in turn will generate calculations and the diagrams that Hawking will eventually memorize. This is the reality that theoreticians—producers of diagrams as well as concepts—construct.[56] Yet, as Hawking's student says, cleverly mixing constructivism, relativism, and realism, "I try *to construct our* universe, *to find a* mathematical *model that represents* our universe. *Ours*. Not the one that *someone has in his head*. Ours, *the one in which we live*."[57]

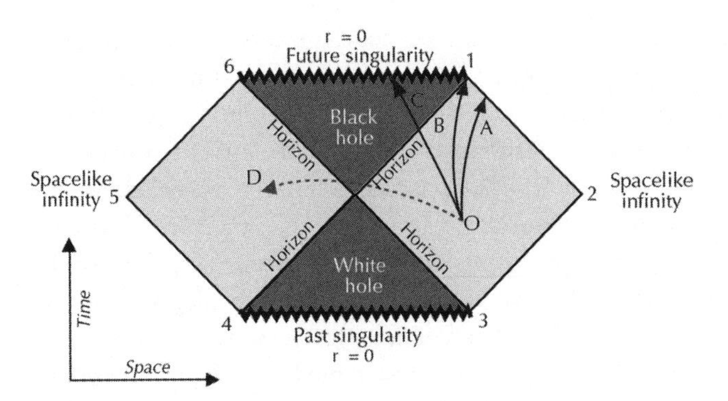

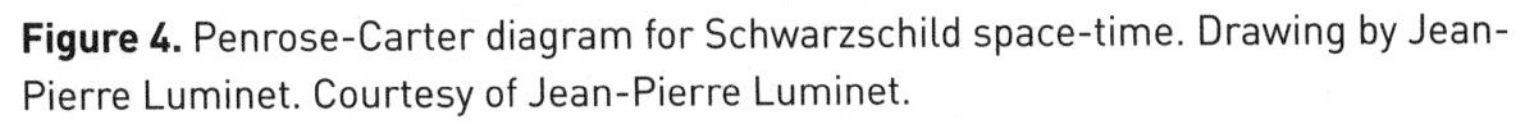
Figure 4. Penrose-Carter diagram for Schwarzschild space-time. Drawing by Jean-Pierre Luminet. Courtesy of Jean-Pierre Luminet.

CONCLUSION

We began with the commonly shared idea that to do theory, one simply has to have a good mind. Because of his condition, Hawking seems to confirm this intuition. This is how he is represented and how he talks about himself. Above all, he is characterized by "his profound thoughts," "his intuition that enables him to go straight to the solution," and "his capacity for visualization." Because he cannot use his hands, he is unable to do the analytic work of laying out each step of a calculation explicitly, line by line. Nor can he use certain diagrams, such as those of Feynman, or a computer to do complex calculations. As we often tend to forget, this type of calculative prowess goes hand in hand with a mobile body that can go to and fro—back and forth—scribble or advance step by step.[58] For Hawking, that would be impossible or would take too much time. As we have seen, the sentence "Hawking cannot go into the details" constantly crops up in his own discourse and that of his assistants, students, and colleagues. Thus, even though his handicap—the immobility of his body—is an obstacle that prevents him from doing calculations by hand or with a computer, it does not limit him in the work of visualization, which, on the contrary, can perhaps be perceived or experienced as a handicap by certain able-bodied physicists. He has, it is said, always been naturally at ease with this work of visualization, pushing it to the extreme because he is unable to use other methods. He has consequently become particularly good in fields that do not involve intense calculative work and in the manipulation of certain problems that he can retranslate so that he can visualize them, that is, *see* either their solution or a path toward them.[59] While this work of visualization is partly introspective (based on the manipulation of mental images), it is also partly based on the use of diagrams—diagrams that he cannot draw himself. This entails the mobilization of his students, who have to familiarize themselves with his language. As his assistants, they learn to communicate with him through *body language* (by asking him questions that he answers with either a yes or a no) and by reading the statements written on his computer or pronounced by his synthesizer, which they interpret and complete. But they also have to know how to calculate and draw the diagrams from which Hawking will be able to select certain information.

This apprenticeship in diagrams is a standard exercise for those who follow the classes of the DAMTP, and the Carter-Penrose diagram is a tool used extensively in the physics of gravitation. However, the difficulty stems from the fact that this diagram has to be adapted to new problems. Students are therefore asked to do the calculations for a mathematical solution of which the diagram will be a projection. They calculate. They draw.

And each time they come to him to present a new problem, they anticipate his questions, putting the corresponding diagram before his eyes. Thought experiments are therefore partially based on—or reactivated by—the use of an exteriorized device, mediated by a student. We are thus looking at a collective thought experiment. Hawking, accompanied by his student and his diagrams, materializes and makes a dimension of mathematical practice visible. As Charles Sanders Peirce comments,

> It is a familiar experience to every human being to wish for something beyond his present means, and to follow that wish by the question, "Should I wish for that thing just the same, if I had ample means to gratify it?" To answer that question, he searches his heart, and in so doing makes what I term an abstractive observation. *He makes in his imagination a sort of skeleton diagram, or outline sketch of himself, considers what modification the hypothetical state of things would require to be made in that picture, and then examines it, that is, observes what he has imagined, to see whether the same ardent desire is there to be discerned.* By such a process, which is at bottom very much like mathematical reasoning, we can reach conclusions as to what *would be true* of signs in all cases [my emphasis].[60]

Rotman, following Peirce's intuition, describes mathematical reasoning as "an irreducibly tripartite activity in which the Person (Dreamer awake) observes the Subject (Dreamer) imagining a proxy—the Agent (Imago)—of him/herself and on the basis of the likeness between Subject and Agent, comes to be persuaded that what the agent experiences is what the Subject would experience were he or she to carry out the un-idealized versions of the activities in question."[61] In the present case, the Agent is sometimes a student.[62] Thus, contrary to what people say about him—and what he himself says (i.e., that "everything is in the head")—Hawking has to delegate more than anyone else. It is his students who calculate and draw. His disability makes visible the particular collective—composed of students, equations, theoretical tools, and representational techniques—without which physicists cannot think.

Conversely, he cannot use paper, blackboard, or computer to draw or write notes to himself—effectively using them as sites of memory. In this sense, his competencies are both more distributed than those of others and less so.[63] As we have seen, the back-and-forth work of scribbling, deleting, annotating, and drafting that accompanies the work of thinking is impossible for him to do. In other words, he follows a hypothesis without being obliged to move back and forth between ideas and various material media; he eliminates the details and finds a means to go straight to the solution, taking shortcuts through diagrams drawn by his students. Thus, he is capable of visualizing, experimenting, projecting himself with his body, identifying himself with objects, and memorizing—*only if everyone does what is required*.

Hawking's condition forces others to do a great deal of exegetical work to make sense of "what he says" and transform it into mathematical language, but it also allows him to insert himself into a context that will enable him to say, "Wouldn't it be interesting if . . . ?" To put it another way, his competencies are more materialized and incorporated than those of others because everything has to be made explicit. Not to enter into details means that the others partially do the work for him by making explicit what he says and by anticipating his request when they propose tools that will enable him to do the work of visualization and to grasp things globally. In this sense, we can say that what we attribute to Hawking's cognitive capacities (the ability to take shortcuts, to see the connections, and to have a "global" point of view) is partly due to the diagrams and what they allow—diagrams that he has partially memorized but that he also has to see drawn before his eyes. Likewise, the student's work of exegesis allows him to keep things implicit. Hawking's "Wouldn't it be interesting if . . . ?" would not be possible without them.

So, if, as Hawking says, it is possible to do physics by using only words to communicate and images to have ideas, to do thought experiments, and to visualize solutions to problems or paths to take to solve them, this is only possible insofar as there is a network of competencies that makes equations and diagrams visible "before his eyes," that translates (transforms) ideas into calculations, and that does the work of demonstration (proof). This network enables him to take shortcuts and literally to see "through calculations." Sometimes these shortcuts will be transformed by his students into calculations, drawings, or diagrams that he will be able to memorize and add to the set of tools that he uses.

Just as the anthropologist's text has made it possible once again to visualize a supposedly invisible part of Hawking's body—his laboratory—by making visible the links and practices constituting it (who is who; who talks to whom; who does what, how, and why), the diagram makes it possible to visualize the universe and its causality, that is—and here I use the terms of one of the physicists whom I met—"who has talked to whom, who has talked to what, and why."[64] Better still, the diagram has made it possible to visualize simultaneously the bodies of bodiless objects (e.g., black holes) and the bodies of bodiless subjects (the physicists—Hawking and his students)—bodies whose mobilization, manifestation, and putting into action differ, depending on the type of reasoning that physicists practice (visual or analytical), but which, as I hope to have shown in this chapter, remain indispensable to any theoretical work.

IV The Media

In 1957 in his book *Mythologies*, Roland Barthes dissected, in two short pages, one of our mythical objects: the brain of Einstein. "Paradoxically," he said,

> the greatest intelligence of all provides an image of the most up-to-date machine, the man who is too powerful is removed from psychology, and introduced into a world of robots; as is well known, the supermen of science fiction always have something reified about them. So has Einstein: he is commonly signified by his brain, which is like an object for anthologies, a true museum exhibit. Perhaps because of his mathematical specialization, superman is here divested of every magical character; no diffuse power in him, no mystery other than mechanical: he is a superior, a prodigious organ, but a real, even a physiological one. . . . The mythology of Einstein shows him as a genius so lacking in magic that one speaks about his thought as of a functional labor analogous to the mechanical making of sausages, the grinding of corn or the crushing of ore: he used to produce thought, continuously, as a mill makes flour, and death was above all, for him, the cessation of a localized function: "*the most powerful brain of all has stopped thinking.*" [. . .] What this machine of genius was supposed to produce was equations.[1]

Today, the brain machine no longer seems to be a myth, for there exists a scientist whose neuro-motor functions—such as "speech," "writing," and "body movements"—are partly made possible by a machine: a computer. People say—as they did of Einstein—that he is a genius.

Like Einstein, "Hawking centralizes the power of his knowledge in his brain." Like Einstein, "he revolutionizes the world through the sole strength of his reasoning." And like Einstein, "he is absolutely alone in the difficulty of this undertaking." Hawking, on the other hand, does not think *like* a machine. Rather he cannot think *without* the machine—his computer—to which he is connected or the collective that works day and night to ensure the smooth functioning of his genius. This is where the mystery lies. Einstein's mythology imagined his thinking as purely material: a brain that functioned like a machine. Hawking's mythology, unlike that of Einstein, disembodies his brain, despite its indispensable connection to an extended body of machines and other humans. Yet this man who cannot move a finger without his nurse's help has come to represent the mythical figure of the lone genius who can grasp the ultimate laws of the universe with nothing more than the strength of his reasoning. The present chapter examines the construction of this paradox by looking at another collective at work: the media.[2]

TRANSFORMING THE COLLECTIVE BODY INTO A DISINCORPORATED BRAIN

Act 1: The Press

Where is Hawking? Where does he appear and in what form? The sociocognitive networks or collectivities that I have been at pains to describe in the previous chapters "magically" disappear when Hawking appears in the popular press.[3] All that remains is the image of a singular body, the property of a whole media network that exploits and circulates it, making it visible and mobile because it is designated as immobile, invisible, and unnecessary. In short, it is because it no longer functions that the scientist's body becomes visible. Left credulously to grapple with this dialectic, we glorify him because he has transcended the conditions imposed on him by his own body. At the same time, the prevailing ideology promotes the myth of a scientist without either a body or self-awareness.[4] For the epistemologist, Stephen Hawking is not disabled: he has become a perfect scientist, a man without a voice—or with a voice that speaks from nowhere—a machine, an angel.

Hence the media, together with the scientist himself—we shall see how he allows or prohibits the media from exploiting his body—transform this collective body into a disincorporated brain. The anchoring point around which this discourse turns is his disability: is it despite or because of it that he has become a genius? Some, like the physicist Jeremy Dunning-Davies, attempt to explain away the "Hawking Phenomenon," seeing nothing particularly special about Professor Hawking "except for his handicap, blown out of all proportion by the media which, moreover, places him in the tradition of the Newtons and Einsteins when in fact his theories haven't been proved";[5] or the journalist Arthur Lubow, who writes that Hawking won't revolutionize physics but will leave us the image of his smile.[6] As we shall see, others transform his handicap into a positive element. The body here represents an obstacle in the view of almost everyone except the scientist, who sees himself as an exception and happily agrees with this claim: "At the age of 35, Stephen Hawking, despite severe physical handicaps, has established himself as one of the world's leading theoretical physicists."[7] Other accounts treat Hawking's handicap not as an obstacle but as the source of his performance and creativity. Since he is no longer distracted by the daily and worldly occupations shared by the rest of humanity, he can devote himself entirely to thinking. He has become a pure cerebral being communicating with the great universal.[8] People are impressed by his exceptional memory and compare him to a Mozart composing a symphony in his head. A journalist, for instance, says, "Anyone who saw the lines of complex mathematics covering the blackboard like musical staves at a recent seminar would have appreciated the comparison."[9]

Stephen Hawking—not simply following in the tradition of Galileo, Newton, and Einstein—possesses (in his head alone) the cumulative knowledge of history's greatest minds. Hawking becomes here almost incorporeal; as one journalist notes, his head is supported by an almost nonexistent body.[10] Finally, Stephen Hawking is seen through the subject of his research:

> There is a phenomenon of relativity known as time dilation, in which time appears to slow down almost to a stop for bodies that approach the speed of light. Hawking alludes, in his book, to what it might be like for an astronaut as he accelerates toward a black hole, as all eternity passes by outside in an instant of his time. *There is a sense in which Hawking himself has experienced a kind of dilation, an inexplicable slowing of a natural process that has added decades to his expected life span.* The answer he seeks may be almost within his sight. But so, perhaps, is the event horizon.[11]

Thus, we witness a process of construction of a genius around the mind/body dichotomy and a transformation of the person of Stephen Hawking into a pure, bodiless, cerebral subject, which we end up reading through his research object. We see the link gradually forming between Hawking's frail body, the disincorporated brain, and the content of the physical theories on which he works.[12]

Yet, while "the brilliant Hawking" appears to be the fruit of a collective construction,[13] we shall see how he personally intervenes in this process. The newspapers are fed by, and partly organize themselves around, quotations from the scientist—quotations that resemble one another down to the last word because entire passages of his life story are now stored, as his personal assistant confirmed, in his computer. As we saw in chapter 1, all the answers are ready and waiting to be used. In this sense, talking to Stephen Hawking means speaking to his computer, which provides a stereotyped version of his life. The scientist's autobiography is henceforth virtually stable.

Though his mother thinks that he does not feel fundamentally different from others,[14] his ex-wife believes that one of her main occupations was to remind him that he is not God.[15] Between the two extremes, we can see how he places himself in the tradition of history's greatest scientists. "Born on Christmas Day 1642—in the year of Galileo's death—Newton prepared for himself a symbolic place in both the Christian calendar and the hagiography of the Enlightenment," as Richard Yeo has noted.[16] Like Newton, Hawking consciously embraces what Stephen Greenblatt has called self-fashioning: "The power to impose a shape upon oneself is an aspect of the more general power to control identity—that of others at least as often as one's own."[17] Mario Biagioli adds an important qualification of this in his study of Galileo: "By emphasizing [this] process of self-fashioning, I don't assume either an already existing 'Galileo' who deploys different tactics in different environments and yet remains always 'true to himself,' nor a Galileo who is passively shaped by the context that envelops him. Rather, I want to emphasize how he used the resources he perceived in the surrounding environment to *construct* a new socio-professional identity for himself, to put forward a new natural philosophy and to develop a courtly audience for it."[18]

We see Hawking employing a similar strategy of self-fashioning by inserting himself in the tradition of the greatest scientists. As he likes to emphasize—and has included in his autobiography on the Internet—he was born exactly three hundred years after the death of Galileo. This statement, first found in a talk given in 1987, appeared in the following form: "I was born on January 8, 1942, three hundred years to the day after the death of Galileo. However, I estimate that about two hundred thousand other babies were also born that day. I don't know whether any of them were later interested in astronomy."[19] This anecdote was also quoted verbatim in an interview granted to *Playboy* magazine in 1990, entitled "Candid Conversation." (We note in passing the effect produced by the discourse of this genius without a body framed by the inviting photo of a curvaceous blonde.) Hence, the quotation was recontextualized, one might say, so that by highlighting all its implications, the journalist could reinforce the filiation between Hawking and Galileo and simultaneously the position of the scientist and the impact of his theories:

> J: Can you tell us a little about your early life, before the secrets of the universe caught your interest?
>
> SH: Yes. I was born on January 8, 1942, three hundred years to the day after the death of Galileo. . . .
>
> J: Galileo was tried and imprisoned for heresy by the Catholic Church for his theories of the universe. Did he have something in common with you?
>
> SH: Yes. However, I estimate that about two hundred thousand other babies were also born on that date. [*Smiles*] And I don't know if any of them were later interested in astronomy.[20]

Bernard Carr, one of his former students, also likes to emphasize this sense of affinity that Hawking has for Galileo. Carr flew with him to Rome when Hawking was to receive the Pius XII medal awarded to a "young scientist for remarkable work," presented directly by Pope Paul VI.

> I remember, when we went to the Vatican, he was very keen to go into the archives and see the document which was supposed to be Galileo's recantation, when he was put under pressure by the Church to recant on his theory that the Earth went around the Sun. . . . I think it gave us some pleasure that the Church finally announced that they'd made a mistake with Galileo, and that in fact Galileo was right. But whether or not the Pope would have approved of what Stephen had discovered, if he actually had understood it, I am not quite sure.[21]

This is also the opinion of Stephen Hawking. He notes that if the hypothesis that he formulated with Jim Hartle of the University of California in 1982–1983—on the absence of an edge for calculating the state of the universe in the framework of the quantum theory of the universe—were correct, there would be no singularities, and the laws of science could be applied everywhere, including at the beginning of the universe. This would mean that he had succeeded in fulfilling his ambition to discover the origins of the universe and would thereby violate the Pope's 1981 prohibition concerning research on the Big Bang itself, because it was the moment of creation and thus the work of God.[22]

Thus Hawking came to write his own epitaph: born three hundred years after the death of Galileo, the last visible trace of his "living" body is appended in the form of a signature in the "golden book" that immortalized him and introduced him into the pantheon of the great. This was in 1979, when he was installed as Lucasian Professor of Mathematics, a chair formerly occupied (as he likes to recall) by Newton—this was the last time he signed his name.[23] "In 1979 I was elected Lucasian Professor of Mathematics. This is the same Chair once held by Isaac Newton. They have a big book which every university teaching officer is supposed to sign. After I had been Lucasian Professor for more than a year, they realized I had never signed. So they brought the book to my office and I signed with some difficulty. That was the last time I signed my name."[24]

The presentation of his sick body is the prism through which the process of discovery appears, making the link between himself and his idea explicit:

> One evening shortly after the birth of my daughter, Lucy, I started to think about black holes as I was getting into bed. *My disability made this rather a slow process, so I had plenty of time*. Suddenly I realized that the area of the event horizon always increases with time. I was so excited with my discovery that I didn't get much sleep that night. The increase in the area of the event horizon suggested that a black hole had a quantity called entropy, which measured the amount of disorder it contained; and if it had an entropy, it must have a temperature. However, if you heat up a poker in the fire, it glows red-hot and emits radiation. But a black hole cannot emit radiation, because nothing can escape from a black hole.[25]

The disease is not mentioned in John Boslough's account of the discovery in his 1985 biography of Hawking.[26] But after publication of *A Brief History of Time* in 1988, Hawking's account was to become the established version.[27] We see how Hawking is again going to be transformed into a subject without a body. For example, in *Introducing Stephen Hawking*, the first two sentences of the passage quoted above are used again, almost word for word: "One evening in November 1970, shortly after the birth of my daughter, Lucy, I started to think about black holes as I was getting into bed. My disability makes this rather a slow process, so I had plenty of time."[28] Except this time, McEvoy adds, "As with Einstein's 'happiest thought,' Hawking too can remember exactly what he was doing when the germ of his idea came to him. [Here there is a drawing of Hawking going to bed.] *He saw in a flash that the surface area of a black hole can never decrease. . . . He did not need paper and pen, nor a computer—the pictures were in his head.*"[29]

Finally, Hawking's handicap and the choice of his research subject are explicitly linked. In the introduction to *A Brief History of Time*, he emphasized that, although he had the misfortune of having a motor-neuron disease, he had been lucky in every other respect—most notably in his choice of theoretical physics, because everything is in the mind, which is why his infirmity had not been a serious handicap.[30] Thus, when a journalist asked him if his illness affected his choice of work, he answered, "*Not really, I had decided to work in this field before I knew*. The only thing it may have affected is that I avoid problems with a lot of equations because I cannot easily write them down. I have to look for shortcuts."[31] Two years later, to essentially the same question—"Why did you choose theoretical physics for your research?"—he answered,

> *Because of my disease. I chose my field because I knew I had ALS*. Cosmology, unlike many other disciplines, does not require lecturing. It was a fortunate choice, because it was one of the few areas in which my speech disability was not a serious handicap. I was also fortunate that when I started my research, in 1962, general relativity and cosmology were underdeveloped fields, with little competition, so my disease would not be a serious impediment. There were lots of exciting discoveries to be made, and not many people to make them. Nowadays, there is much more competition [smiles].[32]

For reasons that were partly practical, the disease was to become a main factor motivating his choice of a research field: "I was lucky to have chosen to work in theoretical physics because that was one of the few areas in which my condition would not be a serious handicap."[33]

This version of events is now the established one. Hence, the fact that Stephen Hawking was born three hundred years after Galileo; his reactions when he learned of his illness (he listened to Wagner and got drunk);[34] the essential role played by his wife in his survival; his jokes about his computer, with its American accent; and his account of his discoveries—are all taken as "given" and used over and over again. He himself either plays the game and lets the media exploit his writings or rebels and intervenes in the construction of his own myth:

> J: According to newspaper interviews and a recent *20/20* segment by Hugh Downs on ABC TV, when you got your diagnosis, you simply gave up and went on a drinking binge for a few years.

> SH: It's a good story but it's not true. . . . I took to listening to Wagner, but the reports that I drank heavily are an exaggeration. The trouble is, once one article said it, others copied it because it made a good story. Anything that has appeared in print so many times has to be true.[35]

This statement was, in turn, also taken and reused countless times before finally becoming part of the myth.[36]

He similarly participated in the collective denial of the media construction of his genius around his handicap. "Kevin Berg, a freshman at Seattle Pacific University, asked [him in a conference], 'How does it feel to be labeled the smartest person in the world?' Hawking rapidly picked out words. He spelled out 'media' and "hype,' which were not included in his computer's library of 3,000 words. 'It's very embarrassing,' was his response. 'It's rubbish, just media hype. They just want a hero, and I fill the role model of a disabled genius. At least, I am disabled, but I am no genius.'" This declaration was made to an audience of disabled persons, who were thus elevated to the ranks of potential geniuses, while the prestige of the scientist was simultaneously enhanced. Hawking thoroughly embraces this media transformation of himself, announcing to the very same audience only moments later that "nowadays, muscle power is obsolete. Machines can provide that. What we need is mind power, and disabled people are as good at that as anyone else."[37] So even though Hawking no longer controls his own body, he nevertheless exercises significant control over his image and the role of his body in the construction of his identity. He does so both in his writings (even if he now refuses all written interviews, his quotations are still used) and, above all, in his public appearances, whether before an audience of the disabled, on *Star Trek*, or in a science documentary.

Thus far, I have outlined a few processes that contribute to the fabrication of an identity, an ego, and a subject, through the relationships woven between the scientist and his peers, the scientist and his work, and the scientist and his research subject. This is partly dependent on the way in which he presents himself—that is to say, on whether or not he plays on the relevance of his disability; on how he plays on his choice of venue (*Playboy* or *Star Trek*); and on his choice of a public: scientific programs or disabled persons.[38] It is also partly dependent on moves made in a network that are attributed (in whatever fashion) to Hawking's "intentions."

But does this relatively stable form—this identity constructed with this succession of discourses, representations, and presentations—really have anything in common with "the real, unique Mr. Hawking," the flesh-and-blood person? It seems that getting closer to him does not enable us to grasp him in his singularity; on the contrary, approaching him is to lose him! Indeed, the computer, by mechanizing communication, causes the subject to disappear from the enunciation and to create a new mediation. Professor Hawking is simultaneously silence and speech. He thus transforms himself into the reflection of all projections—as in any human relationship, but magnified. The machine does not dehumanize the man, but it multiplies his subjectivity. One never knows whether one is dealing with a Hawking who is suffering, or bored, or thinking. As Errol Morris, the director of the movie *The Brief History of Time*, reflects, "You never know if Stephen is annoyed by your question and thinks you're an idiot, or is it that I think that Stephen thinks I'm an idiot, or is it a joke, or is he really annoyed but wants me to think it's a joke? . . . It's a hall of mirrors—like your relationship with any other person, but magnified." Or, as Alex Lyons, one of Hawking's research students echoes, Hawking will sometimes leave a seminar at the department in the middle of a presentation. "They never know if it's because he's pissed off and thinks it's a boring talk, or if it's just that he needs some physical thing like suction."[39]

Thus, the more we work from writings or public appearances, the more we have a stable image and a relatively well-defined ego. The closer we get to the scientist's body, the greater access we have to the extension of his distributed body: his assistants, computer, and students. Finally, when we reach the person himself, we believe we have found the man because we are in the presence of his body, but that is when a multiplicity of Hawkings suddenly appears. In short, the closer we are to him, the more we discover a multiplicity of Hawkings, and, conversely, the more we spread our focus to the multiplicity of constructions and representations of him, the more we have a stable ego. This hypothesis, nevertheless, leaves us with an immense doubt. So where are you, Mister Hawking?

Stephen Hawking Live at GR17, Dublin

On 21st July 2004, at around 1:30 p.m., in the Concert Hall of the Royal Dublin Society, Stephen Hawking gave a lecture at a big conference on general relativity—GR17; the event would end at around 2:45 p.m. Hawking had contacted the organizers and asked to be given a slot because he had solved a problem that had been troubling him and other cosmologists for thirty years. The problem was this: A black hole sucks information-bearing entities into it—anything whose matter is organized, such as a star or a human being, is information-bearing. The black hole then evaporates away—it is Hawking who proved that black holes eventually evaporate and disappear. When the black hole disappears, does the information disappear too?

Hawking's view over the last 30 years was that the information *would* disappear. This topic is part of the esoterica of theoretical cosmology, and quite a lot within that field turns on the answer. Hawking's view was shared by the famous astrophysicist Kip Thorne, while another theoretical physicist—John Preskill—believed the opposite. Hawking and Thorne had jointly made a bet with Preskill, the prize to be an encyclopedia "from which information could be extracted easily." Hawking had now decided that he was wrong and had chosen the Dublin conference to concede the bet.

The conference was a big one—around 1,000 delegates. The news of Hawking's appearance had attracted almost every delegate into the hall along with a large caravan of media reporters. The lawn outside the conference hall sported vans with satellite antennae; four rows of seats had been reserved for press.

After ten minutes delay Hawking was wheeled onto the platform, his head lolling in the wheelchair. With him were his attendants. To one side of the stage were Kip Thorne, who would introduce him; John Preskill, who would accept the prize; and a graduate student of Hawking's, who would field technical questions. At the back of the stage, a big screen displayed PowerPoint slides controlled by Hawking; at the end of the talk it would bear the printed text of the bet written in the schoolboy-humor archaic English that has become the tradition: "Whereas Stephen Hawking believes this and that . . ." and so forth. Thorne made the introductions and explained the procedure. There was a hush. The performance began. [. . .]

[Hawking] sat motionless throughout the lecture. At the end of the talk the press were allowed to ask questions which would be vetted by Thorne before being passed to Hawking. The BBC's correspondent asked the first of these questions; it passed muster with Thorne. The question, asked by Palab Ghosh, was "Could you explain what the wider implications of your theory are particularly as regards the nature of space and time and even the origin of the universe?"

It took Hawking a little more than 25 minutes to answer this question, during which a series of technical points were handled by the graduate student, Thorne and Preskill filled in with a cross-talk routine, there were some long silences, and Thorne explained that Hawking was struggling with the technology. The answer that Hawking gave was "Together with the proposal that [inaudible name of person] and I made on the origin of the universe, this result implies that everything in the universe is determined by the laws of science."

One more question from the press was taken; it was asked by the *New Scientist* representative: "If you've been thinking about this problem for thirty years, and are now confident you've solved it, ... what will you be thinking about next, what is the new problem will you be taking on?" Hawking immediately answered this one without significant delay. He said, "I don't know."

By this time many of the audience had left, but the remainder laughed and clapped—desperate to interpret the most slight of utterances as a sign of profound wit. [40]

Act 2: *The Hawking Paradox*: The Making of the Movie

BACKSTAGE

In 2005, Mentorn Films decided to make a documentary to present "Hawking's latest discoveries," which had just been announced at the GR17 in Dublin, along with his work on proof, filmed in real time. William Hicklin, the director, explained, "The original proposal was that we would spend time with [Hawking] as he worked on the new proof, and then eventually, you know, the end of the film *would be this new proof*."[41] BBC Horizon, which Mentorn executive producer Malcolm Clark had just contacted, immediately agreed to commission the film because, as the director said, "They are interested in anything having to do with Stephen Hawking."[42] But they were also excited about the idea of producing a film that would afford access to the man himself, that would show him as he worked—"a sort of behind the scenes look at Stephen ..."—and a film that would in a sense challenge his reputation.[43]

In short, the idea behind the film was to lift the shroud of mystery surrounding the genius by getting closer to him and by following what people said about him. William Hicklin, a freelance producer and director, was asked to produce and direct the movie. Chosen as director after he had become known in the business for his documentaries—one on a NASA mission to Saturn and one on malaria—he said he immediately agreed, partly because it was a film on physics ("they are quite difficult and quite fun creatively") but mainly because he wanted to meet Hawking in person: "The [main] thing was getting to meet Stephen Hawking!" he told me. And when I asked him why he wanted so much to meet him, he was surprised, and mentioned the first Horizon film on Hawking "the genius" that he'd seen as an adolescent:

HM: Why did you want to meet him actually?

WH: Why? [He looks surprised] Uhhmm. . . . Look, I just, you know, since, I don't know when, I probably saw Horizon about Stephen Hawking when I was fourteen years old or even probably earlier than that, so . . . I mean he's always been . . . set up as this great genius. . . .[44]

Knowing nothing about physics, he had to familiarize himself with the subject. He therefore decided to recruit outside consultants. One, George Harris, was a researcher in physics who had studied chemistry at Cambridge as well as quantum mechanics, and the other was a physicist who had already worked on this type of movie. Hicklin then recounted how he began the process of preparing for the film by reading: "So basically George would read things, I would read things, and we'd kind of swap notes, and I can't remember who read what first, . . . but we read various biographies."[45] Continuing the collective exegesis, the director and the physicist also started reading popular books like *The Life of Hawking* by Michael White and John Gribbin.[46] In Hicklin's case, the Internet was also useful, especially the article by John Baez, an American physicist who had a blog about the Hawking information-loss theory:[47] "That was actually quite informative, because that was written in a kind of nontechnical language, but it kind of went some way to critique the ideas, and so that for me was quite useful. I remember that . . . and trying to get my head around what the issues were."[48]

As their research advanced, two ways of seeing the film emerged: one followed Hawking's ascension to the stars; the other introduced his early work, that is, the original formulation of the information-loss paradox and his solution to it. That is why it was necessary to interview the physicists who had worked with him at the beginning of his career. Penrose, Hicklin said, refused to let anyone near him because he was too busy, but important sources of information were Bernard Carr, who wrote his thesis under Hawking in the seventies, and Leonard Susskind, a Stanford physicist working on a new book in which the ongoing debate with Hawking was recapitulated. Lawrence Krauss, professor at Case Western University, was useful because he was one of the non-string-theory people, so he could explain what Hawking was doing. Kip Thorne also supplied useful information on the way Hawking worked, especially on the origin of the mathematical methods that he tried to use to solve the information-loss problem. Thorne recalled, as he always does, and as he also did with me, his friend's unique ability to think geometrically. For Hicklin, "Kip Thorne was the best about this idea that . . . in a way because Hawking couldn't write down equations, that he had to conceptualize things in his mind and ... as a result of that he ... kind of used, I don't think he actually developed the mathematics. I think that was Roger Penrose that had developed the mathematics, but he just became very adept at using those mathematical techniques, which happened to be kind of more graphical."[49]

Hicklin and Horizon were excited by the idea of presenting more critical voices. But even though some physicists readily spoke about the man behind the scenes, they were by no means prepared to do so in front of the camera. Peter Coles, professor at Nottingham University, was an exception. As the director lamented, "It's not always possible to make the film that you had hoped to make. And also sometimes you don't realize exactly ... how you could have made it till after you finished—sometimes, because it's quite a quick turnaround."[50]

Today Hicklin sees two ways of interpreting his movie. The first is to show that Hawking tried to restore the mathematical method that made him famous at the beginning of his career by using it as the solution to the problem that he had posed in the seventies. The second is to see Hawking's very public intervention at the GR17 conference in Dublin as simply an attempt to make a big splash with the physics community and to restore his status as a leader. Hicklin sees this interpretation as rather unfair because

> he's never really disappeared from the world of physics, ... and I don't think he particularly made any grand claims in the last twenty years or so in terms of physics, ... but ... we did find from various people that there's a general kind of grumbling that he's held up as this great genius, whereas, you know, since the ... late seventies or early eighties there's not ... much you can point to which is brilliant in terms of what he's done. Ummm, but then other people would sort of say well, you know, if you have only done one thing in your career that's brilliant, it's enough for most people [laughs].[51]

In the end, however, Hicklin found that a focus on scientific controversies was too technical for the public; he therefore dropped this approach to concentrate instead on the heroic attempt of Stephen Hawking to regain his prestige in the world of physics—a narrative that he felt was easy to understand and more exciting:

> You know the idea that basically in the seventies he was widely regarded as a brilliant physicist, ... but then he became well known as you know in the public's eye, as kind of the voice of physics, if you like, but his reputa-

> tion within the physics community kind of went down because he wasn't actually doing much work. And so I suppose the idea of the film was that at the end of his life he was trying to regain his reputation within the physics community. Now, ummm, whether he's trying to do that or not, I really couldn't really say.[52]

Hicklin ascribes Hawking's fame partly to the physicist's work but also to the attention that the media focus on him because of his disability—attention that other physicists are envious of and criticize. Interestingly, the director doesn't see himself as participating in this process of collective glorification: "But I think it's partly because of his publishing and partly because of the kind of whole media hype thing, which is not in any way of his doing, well . . . I think he benefits from it, and you can't criticize him for doing that, in any way, but I mean he's only benefited from the kind of media hype that has fed off it . . . and his disability and all that."[53]

MEETING HAWKING

In addition to insatiably reading biographies and articles on the Internet, and interviewing those who had had any contact whatsoever with Hawking, whether close or distant, Hicklin and his team consulted the BBC archives, which they could use without any problems with copyright. They utilized some old BBC Horizon movies showing Hawking with his students. One of them was probably the movie that so inspired Hicklin as an adolescent. By reading the same books, using the same pictures, and interviewing the same students, friends, and enemies, the production team carried out the work of constructing and reproducing a Hawking endowed with similar qualities: his determination, his prescience, his way of thinking geometrically, his birth three hundred years after Galileo's death, his wish to be a physicist from a very young age, his disease, the fact that he was given two years to live, the turning of his disease to his advantage, the fact that all the work is done in his head, and so forth. Erving Goffman speaks of the construction of a biography like a candyfloss.[54] Yet, far from being content with discourse produced around and by the man, the director's idea was to "solve the mystery of the genius" and to spend as much time as possible with him. Learning that Hawking was going to Spain for a week, he contacted his personal assistant and asked him if he could go along. The request was granted. But, as for the ethnologist thirsting for information, spending time with Hawking also meant spending time with his entourage—his assistants and students—who interpret his thoughts: "We spent a week with him when he went to Spain, . . . and so that was really a good opportunity to spend time with him, and, I mean, we kind of learned quite a lot *by talking to people around him*—close to him, particularly Christophe, I think. [...] It seemed that ... Christophe was quite close to him, certainly in terms of research, and also of kind of knowing his opinion of various things as well."[55]

They also tried to film him while he was working, especially on the famous Fridays when he meets with his students. But once again, like the ethnologist, the director, endeavouring to solve the mystery of who does what, discovered that the students did most of the work:

> We were trying to get a kind of behind-the-scenes look, ... and I had no idea of how he worked [HM: Me too, I'm really interested in that as well], ... so I just sort of went in with an open mind thinking we'll film whatever happens, ... and then it became quite obvious early on that effectively, you know, *his graduate students do all the actual work in terms of working out* ... they do all the work in terms of hashing the numbers and working through the equations and all of that, and then they bring it to him, and he obviously assesses it and then points them in different directions, and I thought that [this] actually became ... an interesting part of the film.[56]

Christophe, the student with whom Hawking was working during the shooting of the movie, in addition to doing calculations and drawing diagrams, prepared Hawking's talk in Dublin; answered the public's technical questions during the performance; and acted as mediator between Hawking and the BBC by explaining what Hawking was doing before, during, and after the filming. He met Hicklin and his team, without Hawking, five or six times over the course of the project, and received and answered (with Hawking) the questions that they asked.[57] As Christophe said, "They asked me everything."[58]

They met Hawking personally, several times, to ask him specific questions at the beginning of the filming. He was always accompanied by his student Christophe. As Hicklin recounts, "I met Hawking, I think on the three occasions where I *sort of had an interview*, let's say, with him. Christophe was there as well."[59] The director prepared about ten questions. Hawking was still using his switch, but with more and more difficulty; he therefore only answered two of them. Curious and impatient to hear about the progress of the demonstration, the director wanted to know when the article would be finished. The student, in the wings, responded, stressing the fact that it would take years to find mathematical proof. Hawking, however, promised that the paper would be published by the time "the production was going to finish."[60]

Once again Hicklin and his team were going to have to face the problem described by Hawking's personal assistant (PA). Discovering that it was impossible to have an ordinary conversation with Stephen, he realized that they would not be able to interview him in real time. The questions that they wanted to ask him in the film would have to be prepared in advance. As he puts this,

> Basically with Leonard Susskind we probably interviewed him for about two hours, certainly over an hour, and we had a lot of material which we then could choose, but ... you know it was obvious that we couldn't do that with Hawking, because that would involve such an amount of work for him, so what we did do is ... we edited the film to a kind of rough cut so we knew exactly what ... when we would need him within the film and what kinds of things that we, you know, would want him to say.[61]

To save time, they sent the questions to his student by e-mail. As it seemed that the student was working on the proof, they assumed that he would also respond to the questions put to the professor. To their surprise, they realized that Hawking wanted to write the answers himself. We can see how strong his wish was to be part of the construction of his image:

> To be honest, I had assumed that he would just give all the questions to Christophe. Christophe would write them, and then he would give them a nod and send them back, ... but that's wasn't what happened at all. ... I think that Christophe did do some parts of it, ... but I was actually very, very surprised at how controlling [Hawking] is. I mean ... it would be very easy to get an assistant to do it all and him to just say ok, fine, but *he's very keen to put things in his own words*, so it did take an awful lot of time.[62]

Christophe served as the mediator and relay: "I sent [the director] the answers as *we* wrote them ... as *Stephen* wrote them"[63]—answers that were reread and assessed as they arrived on the director's desk:

> I think he wrote some replies, and then ... I think it went around once, possibly twice. We thought some of the replies were too technical, so we asked him if he could put them in plainer English, and I think we sent back some alternatives ... [that] he rejected, so he was very ... So we sent the questions to Christophe, and he worked on them with Stephen. ... So he sent us the reply, as a text file, which we then reviewed and then, in a way, asked for amendments because some of it was too technical. But as I was saying, I thought it would be quicker because I thought I could just do it back and forth with Christophe. But [HM: But Hawking really wanted ...]

> Yeah, he kind of refused. And in actuality, Christophe had very little to do with replying to those particular questions. . . . Which is great, in a way.[64]

The PA, concerned that her boss might be misinterpreted, reminded Hicklin and his associates to make sure that they had clearly understood what Hawking had said. She urged them to be particularly careful not to add their own interpretations to his statements. In this respect, the fact that the answers were written by him, were printed, and were handed to them (and not simply spoken by Hawking's synthesizer) seemed to reassure them, even though the director had already read most of the answers elsewhere: "Some of them were interesting because they were kind of stories that I've heard before and in the . . . books and everything, and some of them actually weren't, which is kind of interesting, because . . . after you've been researching for a while, the same stuff keeps coming up, round-and-round, and you [think] oohhh that's exciting, that's something new, but there's little of it that is different."[65] Hawking's assistants may have put aside some of the answers they considered to be new and interesting, as they were used to doing, to be recycled later. As the graduate assistant (GA) noted, "When [Hawking] says something that sounds very good, we tend to write it down, we tend to store it somewhere, and whenever journalists ask for comments, and that looks relevant, we tend to give them that. Um, but, he doesn't tend to save everything he says, just because, I suppose, it's not in human nature."[66] Even if he is constantly connected to a computer, saved or not, new or not, Hawking will effectively rewrite all the answers from scratch.[67] His student confirms this: "Even if it seems the same, it's not exactly the same thing. And then the technicians already have a program on him. And if you reuse everything you've written, it's not so good."[68] These answers, in turn, as we shall see, were carefully selected and edited during the filming.

SHOOTING AND EDITING: CONSTRUCTION OF THE INTERVIEWEE

Hawking forces others to adjust to his conditions. In order to facilitate his appearance, everything will be planned so that he can occupy the space designed in advance for him. But once again, just as it is impossible to interview him in real time, and it was necessary to prepare all the questions and answers in advance, so too the way he is filmed has to be adapted: "I think partly because, normally, when somebody is talking, and you've got an interview, then you're looking at their face, and you're looking at the expressions on their face when they're talking, whereas obviously, with Hawking, he doesn't."[69] As the computer screen scrolls down before his eyes, offering a choice of words because he is unable to move either mouse or keyboard, the camera tracks around him, creating the impression that he is moving: "We already previously recorded well (*a*) in the studio and (*b*) in his office . . . some shots where . . . we just used tracks . . . to get some movement in the shots to make it look more interesting."[70]

For the sake of convenience, and to silence the steps of those who are setting up the scene, the interview consists of two phases. The sound and images are recorded separately. The silent man looking toward the camera as if he were giving an interview is to be filmed first, with the sound added later to give the impression that the answers are coming out of his mouth:[71] "In order to make the shot more interesting, then, we obviously have to move the camera, and we didn't want to do *that* at the same time as he was giving a reply because that would . . . wouldn't necessarily work, you know, if we're moving the camera and making noise and whatever. It was just more efficient to get those shots *and then to record the voice separately* . . . because obviously his face largely doesn't move; then you can sort of place the replies to his question."[72]

The décor was chosen with precision. It is reminiscent of Errol Morris's movie *The Brief History of Time*, where Hawking was filmed against a blue background so that the director could project the images onto any backdrop he chose. As Morris explained, "I can place Stephen Hawking where he belongs, in a mental landscape rather than a real one."[73] Hicklin likewise decided to choose an entirely empty white studio, without any objects or humans, an ethereal space where Hawking would give the impression of floating in the air like an angel. Hicklin recounts that "Hawking was a little dubious about it because it might make him look like he was dead, like he was in heaven."[74]

Hawking arrived late for the filming. The bright daylight was replaced by electric lights that the director said did not produce the desired effect. He arrived too late, and he left too early—we can see the traces of his presence left like imprints on the film: "We had chosen a quite beautiful studio; . . . it had lots of beautiful daylight. It was fantastic when we arrived, but unfortunately he was two and a half hours late, so by the time he arrived, it was getting dark, so we had to use a lot of artificial light, so the effect wasn't quite so nice. . . . But we did do some stuff with the . . . quite technical filming, which took a lot of time, and I think by the end of it he kind of had enough so . . . yeah . . . he left."[75]

But before pursuing him, let's rewind the film for a few minutes before he departs, this time to follow the movements of the camera. From afar, we see Hawking in his wheelchair, around which we turn like the earth around the sun—the required conditions having been created to efface both those assisting him and those filming him. He is indeed alone in an empty room.

From close up we see nothing but an eye; the camera has just dipped down and so have we, along with it, on the part of his body chosen to illustrate the mind without a body—the genius: "Partly because there is this fascination with the mind of Hawking and his genius and all that, so we did some very close stuff on his eye, which was very powerful."[76] As if getting ever closer to him could afford us access to the functioning of his brain and the universe—swallowed up by Hawking's blue eye that has become a black hole, it's now we who float in the air among the stars.

But, just as Hawking wants to (partly) control the answers to the questions that the director puts to him, so too—as we saw earlier, and as we will see again now—he also wants to control the image of his body:

> I think he was very dubious about it, and he wanted to see it, actually, . . . and he wanted to know what we were doing at all times, and he did say no to quite a few things, so he was in control. [HM: Like what?] Like certain angles, you know or being too close, and the one with the eye he was very uncomfortable with, and I think we kind of slightly—he probably wouldn't have wanted us to do it, but actually I don't think in the final film, I think it's quite a beautiful shot, so I think he didn't have a problem with it.[77]

In the last scene of the film, we see Hawking working with his student. Here he uses his switch, probably for the last time. During the filming, the switch was discontinued, and the infrared blink switch attached to his glasses was adopted. He nevertheless refused to be filmed with this new system. The eye as a symbol, yes; the eye as a communication tool, no. Even though, in the end, the camera seems to have exceeded the limits of the actor's comfort and will—the fact that he made no objection during the projection of the movie was nevertheless interpreted as an approval—he'll refuse this focus on his eye, while selecting the words on the computer, with a wink of his eye:

> HM: In the movie you see Hawking using his eye, and you said that he was still using his finger, so I was wondering . . .
>
> WH: [The director seems surprised] But we don't see him using his eye! He refused to let us film him using his blink switch.
>
> HM: He didn't do that in the last . . .
>
> WH: No, no, in the last meeting we had with him he was using it, but he refused to let us film it, 'cause he hadn't decided that he was going to go ahead with it yet.[78]

A few frames later, Hawking's eye imprinted in our heads, the student's hand is feverishly writing a series of mathematical equations, reminding the spectator of the complex and hierarchical world of physics:

> We shot one scene, which I think was a Thursday or a Friday lunchtime meeting he had with his students; we shot that; and then we shot another thing, which was specifically with Christophe explaining to Stephen, that, to be honest, is something that we set up. So he did some explaining, and Stephen did some replies; and then afterwards we did some kind of special filming on the board with a special camera, you know, to get very close to it, and because we were so close, we were actually asking Christophe to kind of be writing—following the chalk as he went along; that was more set up to give us some visual material for the film. . . . The equation kind of stands in for a general physics process rather than actually explaining the equations. . . . Well, I had hoped to use, explain the equations, but I think *in the end the equation ended up being more a kind of visual metaphor.*[79]

A visual metaphor like Hawking's eye, as a means of making his genius visible. As the images have already been recorded, all that remains is to get Hawking to talk. For the sake of authenticity, they tried to find "his voice" on the Internet. For the voice is the man, even in the extreme case of pronouncing words written by someone else. However, Hicklin isn't able to locate one:

> So we did try to put his voice in, and so we did look on the Internet to try to find a kind of Stephen Hawking sound-alike, but we couldn't actually find one. [HM: No?] No [laughter]. [HM: I'm surprised, actually, because it's so similar . . .] It's, yeah, I know, I know, there's one on the Apple Mac computer that sounds very similar, but actually it's not the same,[80] ummm, so . . . we did think, you know, 'cause, we already had recorded the pictures with him, *we thought we could just make him say anything* . . . we were just thinking, . . . but we actually, we didn't try to do that in the film, but we just tried to do that out of curiosity, and it wasn't really possible.[81]

To have—to possess—"the voice" of Hawking (could anyone really tell the difference?), the answers had to come from "his" computer, which is to say, to be activated by him or by someone else:

> Basically it has to go out of his loudspeaker, and he has to be there to run it; and in a way I suppose that . . . gives him ownership of his voice, you know, it means . . . that somebody else couldn't kind of use his computer

> and kind of . . . [HM: Yes, I was wondering about that if someone could, or his graduate student, for example, could play the responses . . .] . . . no, no, I mean, . . . it has to come from that computer that's on his wheelchair, so presumably somebody else could use that same computer, but what I'm saying, it's not possible to kind of just get any computer program to get his voice out of it.[82]

The director and the sound engineer were informed that it would be impossible to send the audiotapes by mail,[83] so when the answers were ready, they came from London to Cambridge to record them. "[Hawking] played out the answers to us, and we recorded them."[84]

It was during this recording session that I first met William Hicklin. I learned that Hawking had gone to bed very late to finish preparing his answers. While I was waiting at the entrance to his office, the professor was busy making some changes suggested by his student and altering the spelling of certain words so that the synthesizer would pronounce them correctly. "We finished the answers that morning, and afterwards he listened to them. And there were things that didn't sound right. Like Galileo, if you write GALILEO, it's pronounced GAL(AI)LEO by his computer. . . . So if I wrote like that without paying attention, or if he wrote like that without paying attention . . . , he'd listened to it and want to change it, and it takes ages."[85]

They say the director asked him afterwards to make some changes. Apparently he refused to alter his explanation on the way that he'd found the evaporation of black holes. This answer is not to appear in the final version of the film. His student wants to take a step forward to help him but understands that Hawking doesn't want him to. He tells me that the scientist refuses help in public.[86] That's why he wants to keep Equalizer, so that no one can answer for him. While the sound engineer records Hawking's voice, the director films his computer screen. According to one of Hawking's GAs, "It's also a sort of icon that the world recognizes, those green words flashing across the screen. Ummm, they're not as famous as his voice, but they're still an important part of what he looks like."[87] There I see a citation that I've already seen thousands of times before, although slightly embellished: it's no longer two hundred thousand but four hundred thousand babies that were born on the same day as he was: "I was born on the eighth of January, 1942, three hundred years to the day after the death of Galileo. However, I reckon 400,000 other babies were born that day. So I wasn't that special. Still, I have a strong sense of identity with Galileo. When I was in Rome to receive a medal from the Pope, I insisted they showed me the record of the trial. Shortly after that, the church rehabilitated Galileo."[88]

Before I leave, the director gives me a copy of the answers that were sent to him by mail. In addition to the first one that I just heard, there is another with a familiar refrain: "Like Galileo, I have wanted to work out my own understanding of the universe. Science is based on reason and observation, rather than being handed down by authority from on high. That was the reason for the conflict between Galileo and the church." In the final version, the many babies born on the same day as Hawking will have been edited. A new citation is born:

> Narrator: Stephen Hawking was born on the 8th of January, 1942.
>
> SH: *I was born three hundred years to the day, after the death of Galileo. Like Galileo, I have wanted to work out my own understanding of the universe.*

Among Hawking's responses I also discover a new statement that will not be kept in the movie: "Modern physics has introduced totally new concepts and changed our notion of reality. We need a new philosophical framework to accommodate them. Our old ideas based on common sense won't work." More often, however, I find a set of relatively standardized citations such as the following.

1. Responses showing the mise-en-scène of the disabled body of the scientist at play in the construction of his intellectual itinerary and his discoveries:

   Before I got pneumonia, I was mainly working on cosmology, the origin of the universe. But for the last thirty years, I have always had the information paradox in mind. I had read a paper by Maldacena, a brilliant Argentinean physicist, and I kept thinking about it while I was in hospital. There wasn't much else to do. I gradually realized that I could extend the ideas in Maldacena's paper, and show how information loss manages both to occur and not to occur at the same time.

2. Commonplace descriptions (that one could find practically anywhere):

   If you watched an astronaut falling into a black hole, you would never actually see him enter the hole. Instead, the astronaut's wristwatch would appear to slow down as he approached the event horizon, the boundary of the black hole, and his image would grow redder and dimmer. The astronaut himself wouldn't notice anything special when he crossed the event horizon, but once inside, he couldn't escape, and would soon be torn apart by tidal forces.[89]

3. An account of the discovery of black holes that radiate energy and Hawking's belief that he could win the Nobel Prize (narratives that can be found in *Black Holes and Baby Universes* on pages 107 and 120, respectively):

   At first, most people couldn't believe anything could come out of a black hole. But when they repeated my calculation, they found the same result. So now everyone believes it, even though it has yet to be confirmed by observation. When it is, I will get the Nobel Prize.

4. A director who embellishes or edits Hawking's narrative:

   What was so wonderful about my discovery that black holes have a temperature was that it all fitted together beautifully. It led to the simple equation, the entropy is a quarter the area of the horizon, in geometrical units. Results like that are so elegant, they must be right.

   The relation between entropy and area connected apparently two unrelated branches of physics: gravity and thermodynamics, the science of heat. It shows there is a deep underlying unity in Nature.

   Director: I was wondering if we could combine the two above answers as follows: "What was so wonderful about my equation was that it all fitted together beautifully. *It was the first time* that a relationship had been found between the physics of gravity and quantum mechanics, and *it is still the only equation* that combines these two fields. Results like that are so elegant, they must be right. It shows that there is a deep underlying unity in Nature."

In the final version, only the last two sentences are kept:

   SH: *Results like that are so elegant, they must be right. It shows that there is a deep underlying unity in Nature.*

5. An account of his unique way of thinking geometrically and his eureka moment while getting into bed:

   I could see that there was not a ~~one to one correspondence~~ [direct relationship] between past and future, in ~~spacetimes like~~ black holes, where time stands still at some place. Other physicists didn't think in this geometrical way, and so didn't appreciate my reasoning.

   My ALS made it increasingly difficult for me to do pen-and-paper calculations. I therefore avoided problems with a lot of equations, and

> concentrated on those that I could visualize geometrically. Fortunately, these are the most interesting problems.[90]
>
> I had a Eureka moment while getting into bed, shortly after the birth of my daughter, Lucy. I realized the area of the event, the boundary of a black hole, would always increase. This is very like entropy, which measures the amount of disorder in a system. Other similarities with entropy were soon discovered. But area could not actually be entropy, because that would imply black holes would emit thermal radiation. But, as thought, nothing could get out of a black hole.

Among these answers, only the last will appear in the final version of the film:

> I have always been interested in how things work. I used to take things apart to see what made them tick, but I wasn't so good at putting them back together again. But I think most children are like that. They ask how things do what they do and why. Understanding means that in a sense, you are in control. From the age of 14, I wanted to do physics, because it was the most fundamental of the sciences.

The sentence that put him on the same footing as so many others will have been edited. He is no longer like the others; he is unique. But because of this, the original will become a copy again. And indeed, the statement exists elsewhere. One wonders why it had to "leave Hawking's mouth":

> SH: *I have always been interested in how things work. I used to take things apart to see what made them tick, but I wasn't so good at putting them back together again. Understanding means that, in a sense, you are in control. From the age of 14, I knew I wanted to do physics, because it was the most fundamental of the sciences.*[91]

By talking to those who are close to him and "by reading what Hawking said," the director feels that he knows what Hawking thinks on certain subjects. He continues constructing the biography with the same elements. But what he needs is to understand the Hawking paradox, and only Christophe can explain that to him:

> Well, we sent the questions to Christophe to start with and they went straight to Hawking. . . . I mean he didn't . . . say at any point, we actually didn't ask him really . . . what would Stephen say to this sort of question. I suppose partly through *talking to lots of people who knew him; and just reading stuff that he already presented on the subject, I think we knew what his*

> *opinions were about various ideas*; and so we had a pretty good idea. What we needed Christophe for more than anything else was to try to explain this whole information-loss paradox and what Stephen's new approach was; and . . . one of the big problems for me in making the film was to try to represent those ideas in the film, and I don't think we actually . . . really managed to achieve that, because they are so complicated.[92]

After spending eight weeks researching, two weeks filming, and eight weeks editing, Hicklin says he has, more or less, understood, not the mathematics and physics at play in Hawking's work, but how to talk about concepts. Yet in the end he admits failure in the explanation of Hawking's ideas. The director himself believes that Hawking's ideas don't translate to film—they can't be visualized:

> And I think . . . I don't know, I mean I suspect somehow that there are ideas that are so complicated really . . . that when you try to visualize or try to conceptualize what the ideas actually mean, then . . . there's quite subtle differences between one approach and another approach, and they seem slightly nonsensical to be honest, . . . which you can play with in a film like that; you can sort of make it all sound crazy in this kind of mad world of physics [way] that we're all a hologram or you know, whatever, until you look at what's happening behind the wall—anything could be happening. Although those kind of ideas, you know, are quite good fun, you can play with those initially, but . . . in the end there wasn't any one idea like that that was . . . central to Stephen's . . . new approach . . . so it was difficult to get into the physics of his new idea . . . I suppose.[93]

The only place that he feels he really succeeds is in presenting the role of the students in the way Hawking works: "That's what we were talking to Christophe a lot about, ways to try to . . . achieve that in the film, and I think . . . that we didn't manage to do that very well, and in the end . . . the film was . . . about the process of how Hawking works on his ideas, and [in particular] the idea . . . that he's working [on now] with Christophe . . . we got that I think."[94]

Hicklin is disappointed. And with reason, for the genius is not located in the brain but emerges from this collective. That's what he says constantly, but he doesn't realize that his disappointment is because of this; he wants to ascribe it to the fact that he arrived too late:

> . . . as I was saying, . . . the whole fascination is how [Hawking's] mind works, and . . . I kind of have to say that after working with him I didn't

get much closer to that from being with him *but only from talking to other people about him*. So, you know, because we didn't see much of that process happening other than just the work he was doing with Christophe; which to be honest, ... because Christophe ... was kind of getting to the end of the PhD, ... so most of the ideas and most of the talking were coming from Christophe at that point, so it seemed like the mind of Hawking at that point [HM: Was Christophe!] yeah, yeah [laughter].[95]

Still perplexed, the director adds, "*I don't know how his mind works in any way*. Ummmm, so ... he's working on this problem with Christophe, Christophe will put equations up on the board and explain what he's doing, ... then Hawking will say something [but] *I don't know whether that's because he's got an idea in his head ...*"[96]

Once again Hicklin, by trying to film Hawking busy working with his student, wonders about Hawking's role as a figurehead:

HM: Did you see them really working together, because we see a little in the movie.

WH: There's a little bit, I mean it's not very exciting Yeah. Mainly we filmed a session where Christophe presented him with some work he'd been doing, and Hawking just made a couple of comments.

HM: Do you remember what kind of comments?

WH: No.

HM: I mean it was more elaborate than yes or no, or ...

WH: It was ... but again, it was quite brief. ... It was quite brief. I mean, I can't remember what it was, it was something like "Are you sure?" You know what I mean, it was slightly questioning, and I can't ... but it wasn't a big long technical response. It seems to me partly what Christophe said, he worked with him originally when he started doing his PhD, and there was this problem, and it was just kind of set out, and it took him two years to understand the problem, so I don't know how much Stephen kind of exists or works with his students as kind of a figurehead,[97] and how much it's kind of one on one. You know, it seems to me, being there, his actual interaction with students is obviously quite limited.[98]

Yet, says Hicklin, even though the student performs on his behalf, he remains loyal to his professor, whose name will allow him to explore other worlds:

> But I mean . . . I spoke a lot with Christophe, and I actually met him after the film and everything, and he had plenty of opportunity to say . . . that it's a whole load of rubbish, you know, I do all the work, whatever, and he never once did; he was always very . . . because, you know, we talked about some of the criticisms of Hawking, and he was very, I think, *loyal* is the wrong word, but was very impressed with Hawking, and he was obviously [a] very smart and very capable physicist working at a very high level, and he had a huge amount of respect for Hawking and his ideas, and he was one—more than anyone, I think, [who] knew . . . and worked with him. . . . So you know you always kind of hope in a slightly journalistic way that you'd get there, and you'd, somehow, you'd end up going out for a drink with somebody, and they'd say, "You know what, Hawking this, Hawking that, or we do all the work," and he didn't ever once say that, and . . . he was quite definite about how Hawking helped him, and so, I don't know, . . . sometimes, you go into a place, and you get a feeling that something's not quite right, but it wasn't like that really. What Christophe did say, was that there is kind of a Hawking thing actually what he's doing now, he's making films [laughter] about Hawking with Hawking now, and he just said that the whole Hawking thing has opened up huge doors for him.[99]

Knowing Hawking for Hicklin means knowing those surrounding him. This does not seem dissimilar from an actor-network formulation; that is, Hawking is the product of an association of heterogeneous elements that constitute his body and mind; he is the result of a process of composition; he is indistinguishable from the network of entities on which he acts and that act through him:

> I spent quite a lot of time around him and I think it's very, very hard . . . for anyone to know what he's like as a person, you know what I mean? [HM: Yes.] 'Cause he says so little, but I think just knowing the people who are around him, and who he employs . . . and his . . . graduate assistants and postgraduates, and people, you do get a sense of who he is and then . . . [HM: By talking to them, you mean] . . . talking to them, and talking to them about him as well, and also, just occasionally, he would chip in with a few comments. . . . I mean what was hugely impressive [is] the fact that he just keeps going and keeps going.[100]

Yet this view does not take into account Hawking's resistance to having words put into his mouth. Indeed, Hawking's singularity sometimes emerges when he doesn't answer in the way one expects, that is, as one has seen him doing before:

> We, he went to meet the president of Spain as part of this trip, and . . . Judith had written this very formal kind of introduction saying, "Dear Mr. President, you know what a great pleasure it is to be invited, blah, blah, blah, blah." . . . In the van on the way there, he completely rewrote it, and he just kept saying no, no, no, and all he ended up saying was something like, pleasure to meet you. . . . Your ideas on Iraq are so much more refreshing than Bush or something like that, because he was very antiwar, and . . . he just didn't want to give any sort of, . . . he just wanted to go straight to that point, whether because he knew that journalists were there, and they would pick up on that or whatever I don't know.[101]

Yet this anecdote will not appear in the movie. Only the clichés will be recycled. Faced with the multiplicity of Hawkings that he would have to present, Errol Morris, in his film adaptation of *A Brief History of Time*, concluded that it would only be the creative and courageous scientist that would be remembered: "The documentary provides," in Morris's words, "very little biography, but a biographical sketch that suggests Hawking's persistence, discipline and creativity, following the onset of the progressive illness."[102] Similarly, although Hicklin doesn't see the genius working—even if he presents him as such—he'll remember the man for his determination, his perseverance, and his humour.[103] It is this determination that he wants to highlight in the last part of the movie, where he shows Christophe working with Hawking:

> The overriding thing that I just came away [with] from this was his being incredibly impressive in terms of his perseverance and . . . his stubbornness; . . . just sort of carry on and keep doing what he's doing. . . . It's completely unscientific, but it just feels [like] that's the reason why he's still alive, because he's determined not to give up in a way, because I think if he did, if he did give up, then . . . it would be so easy for him not to continue struggling, because it is a struggle, I think.[104]

The last phase of the movie will be accompanied by some rewriting, following several exchanges between the student, the director, Hawking, George the consultant, and the executive producer. Talking about the student, Hicklin says, "I do think that at the end of the edit there was a certain amount of backwards and forwards with some of the phrases that we used, and, you know, he would say, 'Well that's not right, or this isn't right.' [. . .] I think that Christophe was largely in agreement with what we replied, and Stephen had some issues, I think, which he changed again."[105] It is also through what he has read about Hawking that Hicklin judges whether or not what he said was right:

> As far as between me and the executive producer, who got involved in the end, he would rewrite stuff and say, “Well, surely you could say this because . . .” And I would sort of say, “Well, I don’t think you can, ’cause that means something different in the physics,” and that was just from my understanding of it from however many weeks—eight weeks or whatever, you know—and . . . just because I’d been reading a lot . . . about how other people have explained it, I think you get to understand . . . you can explain concepts. . . . There wasn’t . . . a definite structure in place, if you like, to do that.[106]

And thus the work of collective exegesis continues, participating in the construction of this narrative that has become a laboratory of proof. The film will be shown to Hawking at the end—after all, the team wanted to be sure they’d understood “what Hawking wanted to say.” This was dangerous, for had the film not met with approval, it would have greatly hindered them. The director feels constrained by Hawking’s presence and doesn’t want to displease him, as he senses that the movie is somewhat critical. But with a wink, the movie is approved.[107]

Act 3: *The Hawking Paradox*, Directed by William Hicklin for Mentorn Films (2005)

The narrator’s voice rises. Doubts and questions elaborated earlier vanish, as if by enchantment. Hawking is presented as “the most famous scientist in the world,” “a true celebrity scientist,” “a global personality with appearances across the world,” “he’s become a by-word for genius.” Hawking “says” that he has always been interested in how things work. . . . “He was born three hundred years to the day after the death of Galileo.”

The story, the narrative, circles back upon itself, meanders, and roams but always returns to the notion of the genius. The narrator explains that Hawking made his name almost immediately; working on the Big Bang established him as a great scientist. But mysteries remained. Hawking turned his attention to black holes to shed light on them. A cascade of explanations by Krauss, Thorne, and the narrator allow us to comprehend why Hawking thinks black holes can help him understand the nature of the universe. The narrator mentions that they will also lead to his most controversial idea.

[*Close-up on Hawking's computer.*]

SH: *Black holes are places where space and time come to an end, and matter is crushed out of existence. If we could understand how time comes to an end in black holes, it might help us to understand how time began in the big bang.*

Thorne looks out from the screen and tells us that during the 1970s Hawking was a dominant figure in the field. At the very same time, the narrator mentions that Hawking was diagnosed with ALS and given two years to live. Hawking decided, as he puts it, "to ignore this prediction," turning his disease into an "advantage." Here an extract from a BBC film appears, where one sees Hawking and his students; a close-up on Hawking's eye is accompanied by the now-familiar story told by Kip Thorne, who explains that Hawking's handicap has made it possible for him to develop a unique way of visualizing things in his head and that he is able to do a larger range of calculation than if he hadn't contracted ALS. John Preskill from Caltech then tells us about Hawking's remarkable intuition about spatial relationships. While the camera moves back and forth between Hawking's eye and the black holes, we are told that Hawking's particular way of thinking and seeing the world will enable him—and us—to comprehend black holes better than anyone else. Before him they were no more than mysteries to be solved. During the ten years that passed after his diagnosis, Hawking started understanding better and better the functioning of black holes, but he wanted more, he wanted a complete mathematical description. This is extremely complex, but the narrator explains further.

Narrator: It was now that Hawking's new way of working came into its own. He imagined a series of elaborate interactions, between a black hole and the different forces of nature—from gravity to those that govern the quantum world of subatomic particles. The equations that governed all these different interactions were long and complex. But in one of the great insights of modern physics, he managed, now, to boil them all down into one single equation. [. . .] From particle physics to Newton, everything Hawking knew about a black hole was now brought together in a small, but audacious piece of mathematical brilliance.

Professor Preskill: He was able, just by visualizing the process in his mind, to get to the right answer. It's a marvelous thing.

Narrator: It was a triumph. And it confirmed Hawking's reputation as a genius.

SH: *Results like that are so elegant, they must be right. It shows that there is a deep underlying unity in nature.*

The narrator pursues the story as it becomes more and more complex. Indeed Hawking found that at the root of his discovery lies a paradox that puts the very foundation of physics into question. This discovery initiated a battle that has lasted thirty years.

[*Time shifts forward to the present.*]

Almost alone, Susskind, the narrator tells us, understood the implications of Hawking's discoveries and became his fiercest adversary. A succession of commentaries—between Susskind and the narrator—follow, allowing the viewer to visualize the place where Hawking and Susskind met for the first time at the house of the pop psychologist Werner Erhard.

[*Pictures of individuals screaming and contorting their bodies on the floor.*]

It is precisely there, concludes the narrator, that Hawking introduced his work on the black holes.

Narrator: It soon became clear that his new theory was astonishing—for Hawking said that he could prove that bits of the universe were disappearing. This was a bombshell.

Susskind believed that this violated all the principles of physics: "We stood," he says, "at the blackboard, in this electric moment of stunning confusion."

SH: *Leonard Susskind got very upset. I think he was the only one in the room who fully appreciated the implications of what I had said.*

The narrator tells us that there exists now a "before and an after Hawking." Before him, one believed that black holes were black; after, they could radiate energy.

Peter Krauss, the narrator, and Susskind speak, and we learn that the possible disappearance of the black holes challenges the laws of quantum physics. Indeed if information is lost, it could have the most dramatic effects on the ordinary world. Cut to Hawking:

SH: *Scientists usually assume there is a unique correspondence between the past and the future, cause and effect. But, if information is lost, this is not the case. One wouldn't be able to predict the future with certainty, and one couldn't be sure what happened in the past.*

Susskind tells us that he went back home totally obsessed with the questions that Hawking had just expressed.

The narrator adds that Hawking's announcement coincides with a change in his status. From here, we return to British television, where Hawking is introduced as the most important physicist since Einstein, a genius, a global celebrity, a best-selling author. We even see him meeting the Pope.

Narrator: A man who said, "Science should read the mind of God."

Peter Coles compares Hawking to an oracle. The narrator recalls that while Hawking's celebrity grows, he continues to maintain that the information paradox is true. This is a troubling theory, as the narrator explains.

Narrator: If the Information Paradox was true, then all of these black holes were machines, eating up information. And so, across the world, physicists were increasingly keen to solve it—no one more so, than Leonard Susskind.

We then learn through an exchange of commentaries by the narrator and Susskind how Susskind is going to resolve the problem that obsesses them all by showing why information is not lost in black holes. Maldacena confirms this result in an article entitled "Eternal Black Holes in Anti-de-Sitter."

The balance has shifted; Hawking now stands alone, against everyone.

Narrator: The tide was turning, again, against Hawking. The growing consensus was that he just *had* to be wrong. Cause and effect were related, after all. Our memories were safe. But one man begged to differ.

We return to Hawking, where we learn that to stay alive, he has to rely on his wife and his nurses. However, though weak and alone, the hero returns to mount his attack. Kip Thorne testifies that Hawking's disease weakens him more and more. The narrator adds that Hawking "can't work alone," and he has to recruit a young researcher to help him. The student tells us that it is a dream to work with such an icon and, as he told me previously, that he works with him using words more than equations. He is represented here as Hawking's amanuensis; he fleshes out his ideas, he does the work of proof, and he struggles to decipher the thoughts of the genius. Then Hawking decides to tackle the paper by Maldacena: "Stephen asked me to ... have a look at that paper. So, I took a little while, to read it, a little while being a year and a half. It took me a while to understand it." He is making no headway. But a tragedy occurs. Pneumonia devastates Hawking. He is brought to the emergency room. His body is weakened, but his mind is already somewhere else. This is when, we are told, he has his idea—the Idea.

Narrator: Hawking's body was now seriously weakened. But his mind was elsewhere. Throughout his career, Hawking had wrestled with the infinite—the vastness of the universe, the beginning of time itself. But now, on what appeared to be his deathbed, he returned to his favorite field, the black hole. For Hawking, this ultimate destructive force of nature was familiar territory. He had circled its horizon, and plumbed its depths. But this time, as he contemplated it, he felt that he could see something new. For the first time in thirty years, he could see a fresh way to consider the black holes' greatest puzzle—the Information Paradox. A new idea, which, if true, would confirm his position as the world's foremost expert on black holes. Once again, Hawking defied his doctor's predictions, and within three months he was discharged from hospital and working on his new idea.

A close-up on Hawking's head follows, then on his eye, followed by the representations of the universe; then we return to Hawking floating in his white studio. Christophe mentioned that when Hawking came back, they worked day and night. The narrator tells us that Hawking wants to announce his discovery at the most prestigious physics conference. Garry Horowitz mentions that with someone of Hawking's stature, he couldn't do anything else but accommodate him. And we are back where we began in act 2 of the present chapter, at the GR17 conference in Dublin.

[*Pictures of Hawking at the conference.*]

SH: *Can you hear me? I want to report that I think I have solved the major problem in theoretical physics that has been around since thirty years ago.*

[*Close-up on Hawking followed by representations of the universe.*]

Narrator: Hawking's speech turned out to be one of the great U-turns of science, for information, he now admitted, was not lost in black holes after all. The idea he defended for thirty years had been wrong, all along. But after the shock, there was a twist in the tale. Hawking claimed Susskind wasn't right, either. And instead, he now had his own solution, one that was to leave his audience largely bemused. It was based on the familiar theory that the universe that we live in might be one of an infinite number of universes, each with its own different history. In some, a black hole would exist. In other universes, it would not. To understand the real effect of a black hole, you had to combine all of the parallel universes together.

[*Pictures of Hawking and his head.*]

SH: *One, therefore, has to sum all the alternative histories with and without a black hole. Information is lost in the black hole histories, but information is preserved in histories without a black hole.*

[*Images representing black holes.*]

Narrator: In effect, those universes where black holes existed would be canceled out by those where they didn't. And that meant information didn't disappear, because there would be no black hole for it to become trapped in, in the first place.

SH: *If one waits long enough, only the histories without a black hole will be significant, so in the end, information is preserved.*

After the conference is over, the critical voices of Susskind, the narrator, and Coles take the stage. The film's final scene takes place in Hawking's house. Christophe sits near him. Two keyboards connect them to the screen. Christophe tries to decipher the professor's thoughts. He looks at him with anxiety. One observes a dialogue in which only one person speaks.

Christophe: A *t*?

He understands that he's wrong.

Christophe: An *f*?

Christophe: Infinity? Infinite? [*click*]

Narrator: So will this latest idea of Hawking's turn out to be a fitting coda for a lifetime of achievement? That will depend on the work he is doing right now.

Christophe: Well . . . defined?

Narrator: He is trying to flesh out his idea, with a mathematical proof. If it is to convince his critics, it will have to be just as brilliant as his best work.

Christophe: Was it defined? No? Sorry.

Narrator: Progress is tortuously slow.

Christophe: Described?

Narrator: Hawking now finds it hard to spell out words. And so Christophe tries to anticipate his professor's thoughts.

Christophe: Do you want an s?

Narrator: Hawking guides him with small movements of his face.

Christophe: Do you want an s? No? Sorry. A *d*? A *t*? An *f*? Formulated?

Prof. Leonard Susskind: I don't know the details of his proposal, and so I can't really comment on the, on how successful it will be. I can't say, at the moment, whether Stephen's new ideas will materialize into anything interesting.

Prof. Lawrence Krauss: Stephen posed this incredible problem, and, what could be better, after having done that, than to bring it full circle, and solve the very problem he posed.

SH: *I have no intention of stopping anytime soon. I want to understand the universe and answer the big questions. That is what keeps me going.*

[*End (Credits)*]
Executive producer, Malcolm Clark, editor, Andrew Cohen

Figure 5. Thumbnails of the BBC Horizon documentary *The Hawking Paradox*. Courtesy of the BBC.

CONCLUSION: FROM THE OFFICE TO THE SCREEN: *THE HAWKING PARADOX*

Recall that the aim of the movie was to enable the viewer to follow Hawking as he works to prove his discoveries. It was also intended to put his reputation into question. In the end, the director never saw the genius working, at least not the one he had heard of and that he had seen in movies similar to the one he was about to make. Indeed, after getting close to him, he even doubted that the genius really existed. Nevertheless, it was Hawking the genius who was to be reproduced in the end. Between Hawking agreeing to be filmed and HAWKING on the screen, we have followed a number of steps.

We have seen how the director and his team pursued the work of collective exegesis described in the previous chapters. To make the movie, they first delved into books and articles on the Internet explaining Hawking's work. They then watched documentaries on him, recycling images already produced by the BBC. Not content to limit themselves to books or films, they wanted to get even closer. But getting closer to the scientist meant talking to colleagues, friends, and enemies, who were enacting a similar process of exegesis by interpreting Hawking's work and reproducing a number of standardized qualities. Even the critiques were similar, as, for example, when reference was made to his being a kind of oracle. Then, the director and his team wanted to move a step closer, but this time it was the assistants who did the exegetical work. Thus, for example, they advised the team to make sure to understand clearly "what Hawking was saying" and not to add their own interpretations to his thoughts. It was Hawking's student, of course, who played the most important role in explicating his ideas, especially since the work that was the subject of the film was new and had not therefore been recycled much. Finally, they gained access to the man himself but said that it was difficult to know him as a person; thus, by the end of this process, the director concluded that he could only know Hawking the genius through the reconstruction of the network to which he was attached. This movement echoes what I showed at the end of the first part of this chapter. The director's description actually seems to correspond to that of the analyst. The more he works from writings or programs, the more stable are the image and the relatively well-defined ego that he finds. The closer he gets to the scientist's body, the greater his access is to the extension of his distributed body—his assistants, computer, and students. Finally, when he reaches the person himself, he believes he has grasped an individual because he is in the presence of his body, but that is when a multiplicity of Hawkings suddenly appears—a multiplicity of Hawkings because Hicklin can't grasp who he is—despite recognizing his will and determination. We discover with him that not only is it difficult to know Hawking when one gets close to him, but also that he is even more distributed than we had thought. It will be necessary to film his computer with his answers, on the one hand, his face and his smile, on the other, and then record his voice afterward. In fact, Hawking as a person, it seems, is going to be reconstructed piece by piece. Yet the camera and the director move at cross-purposes. Such is the *Hawking Paradox*: during the interview, the director tells us that the closer we get to the man, the more we lose the genius; the camera, however, tells us that the closer we get to the man, the more we grasp the genius. We see the man sitting in his office; then the office disappears, as well as the books, the articles, the blackboard, and the students. He is nothing more than a body floating in a white studio. Then the body disappears: he is nothing more than a head. Finally, the head disappears: he is nothing more than an eye. Then the eye disappears: we are in the universe. All the mediations—performed by those surrounding him and those filming him—vanish; Hawking is no more than an eye, a transparent window onto the world. And there, despite his doubts, the director who becomes a narrator presents the great mind, alone against all, who transcends his body and displaces himself in the universe. Even the student—who, as we now know does so much—appears as little more than an acolyte who fleshes out the master's thoughts.

But there is another paradox still: on the one hand, the director says he wants to know Hawking and get closer to him; on the other, he already seems to know what he wants to show of him and gives the impression that he could have made the movie without him. After reading the texts on and by him, he knows in advance when he's going to speak and what he wants him to say; he says that he knows what he thinks; he expects the student to answer the questions; he looks for Hawking's voice on the Internet; he's going to use known answers rather than presenting, for example, the singularity of Hawking rewriting his announcement on the war in Iraq. Hawking constantly tries to intervene to impose his presence, but his interventions often seem to be erased—not only by others, but by himself as well.

We have seen Hawking's role in the collective work of constructing his genius. We see it not only in the wish to make his work public, when he says that his paper will be written on time, or in his attempts to control what he wants to show of himself—he allows himself to be filmed in a white studio, and agrees to play the genius but refuses to show his new communication system and so on.—but also by his desire to intervene in the writing of himself. At the beginning of this chapter, we were confronted by his subjectivity, that is, the transformations in the use he has made of his disease in accounts of himself and in the standardization of his legend. Here, he refuses to delegate the drafting of the answers to his student; he refuses to delegate the work of writing answers to his computer; he rewrites the sentences from scratch—even though certain standardized answers were probably saved on his computer, or even words like *Galileo* for example—which were probably part of his vocabulary. He refuses to delegate to his assistants the task of making his computer talk when the director comes to record his answers. Yet, his resistance to others doing the work for him often seems in vain. For example, many answers, although rewritten from scratch, are similar to those that he has already produced. In other words, they could have been found elsewhere, in the press or in his computer. Moreover, among all the answers that he produced, very few were kept in the final version of the film. Indeed, few of the standardized answers (e.g., those in which Hawking says that he uses neither paper nor pencil but thinks visually, or the description of his discovery after the birth of his daughter Lucy) were of interest to the director. Finally, regarding the standardized responses chosen, it is the director who has the last word on whether or not they will be used. The fact that "Hawking was born three hundred years after the death of Galileo" will be retained, albeit in shortened form. Sometimes Hawking is cut out in the final edit, confining him instead to his legend. For example, though he repeats the story that he was interested in the way things worked from a very young age, while generalizing his position by saying that other children had the same interest, this will be edited. By editing the sentence in which he puts himself on an equal footing with others, he is singularized as a genius. The presence of Hawking rewriting his history, however, is erased, and the citation goes back to being a copy. It could have been found elsewhere.

What the film presents is the genius presenting himself: "*Born three hundred years after Galileo . . .*"; "*I have always been interested in how things work.*" Two other answers by Hawking will be placed just after the announcement of his discoveries, to underscore their importance or veracity: "*Results like that are so elegant they must be right*"; or Hawking referring to Susskind: "*He was the only one in the room who fully appreciated the implications of what I had said.*" A third answer presents Hawking explaining the importance of his discoveries: "*Scientists usually assume . . . But, if information is lost, this is not the case.*" As for the last answer, an attempt is made to explain Hawking's results; they are surrounded by comments from the narrator to make them meaningful: "*One therefore has to sum up all the alternative histories, with and without a black hole.*" The last three citations, although edited, don't seem to have been printed elsewhere. Many articles commenting on the movie's release have already started to recycle them. Indeed, information is never lost. . . . Our memory is safe.

V Reading Hawking's Presence

An Interview with a Self-Effacing Man

In the previous chapter we witnessed a strange paradox: the closer one gets to the man, the more one discovers a multiplicity of Hawkings, while the more one turns to the multiplicity of constructions and representations of him, the more one sees a stable ego appear. In this chapter, I intend to explain this phenomenon by "zooming in" on my subject, recounting my first interview with him. However, rather than focusing only on the content of the interview, I also want to present an ethnographic study of the interview itself. I focus on the computer, on the intelligent system that mediates Hawking's relationships with others (including myself) as well as with himself. In doing so, I try to explain the functioning of one specific aspect of his extended body "in situ": that composed of his assistants and his machines. However, this chapter goes further by addressing four interrelated questions:

1. The notion of *mise-en-présence*, the *to be there* so dear to anthropologists—what does it mean in this particular context? Do we really learn more about a person (in this case Hawking) when we are in his or her presence?[1]
2. Behind this methodological question, there is another: What difference does it make when one deals with texts as opposed to a person?[2] With Hawking, this question becomes even more complicated, for in his presence, one is dealing with a kind of transcript. I will show, then, that far from escaping the world of writing and representations to find ourselves in direct contact with a person, we are instead confronted with an illegible body and a text being written about Hawking, the Lucasian Professor of Mathematics—a text that will be stored in a part of his computer, waiting to be added to other narratives about Hawking.

3. What kind of body is Hawking's body? Indeed, where is his body? I will show how Hawking's competences are exteriorized and incorporated in the human/machine-based network around him, thus giving him the possibility of maintaining his identity as Stephen Hawking the man and STEPHEN HAWKING the genius physicist, while at the same time deleting the presence of his flesh-and-blood body. Following from this, I also try to explain how and why the interview itself became entangled in these processes of embodiment and disembodiment.
4. Finally, what can we say about the role of machines? Though machines or instruments are often represented as transparent extensions of the body (by, for example, Heidegger, Merleau-Ponty, Polanyi), in this case they play a more active role. I describe how the machines and instruments around Hawking disrupt as much as they make possible the conversation (our interaction) while making them disappear at the same time. In this sense, we will see how the relation of power between the "anthropologist" and "the native" shifts. Indeed, the establishment of the "text" is central to what ethnographers do both in the field and thereafter.[3] In this case, however, the "native" writes, registers, and edits the interview, leaving the ethnographer powerless and destabilized.[4]

FIRST CONTACT

June 25, 1998. The man I have been trying to meet for over two years has agreed to meet with me today, one week before I am to leave England. I have found, it seems, the magic word, the key to pass through the gates guarding the Lucasian Professor of Mathematics. Each previous attempt failed. When I first wrote to the professor (then to his personal assistant) as a philosopher of science excited by his discoveries, and then again as a sociologist keeping her ears open to the "media events" surrounding each of his moves, the answer was always the same: *No*. No, because the professor is too busy. No, because the professor thinks that it's not appropriate. The magic phrase—the open sesame—was *Lucasian Professor of Mathematics*. After being commissioned to write an article in a book on the history of the Lucasian professorship, I asked again for an appointment with Hawking.[5] One day later, an e-mail was waiting for me saying, "Professor Hawking would like very much to meet you to discuss the chapter you are going to write in the said book. Could you, please, contact Pr. Hawking's Personal Assistant at the address below. . . . TA, graduate assistant to Professor Hawking."

June 25, 1998, 4:15 p.m. The waiting room of the Department of Applied

Mathematics and Theoretical Physics of Cambridge is swarming with students. It is teatime. Tom invites me to go into his office—a small office, almost empty, with a big table with two big black boxes on the left side and a laptop and a desktop computer on the right. A drawing of a wheelchair attached to a computer hangs on the wall. Tom explains to me that a "tour" has been organized for me with Stuart Rankin, the department computer officer, and Paul Shellard (a former student of Hawking's who now collaborates with him and takes care of the administrative part of the department). I just came to see Hawking and his assistant Tom; I didn't expect this "tour." I was also surprised when Tom told me that if I wanted, I could interview Hawking's nurse, Christine, as well. The "technicians," apparently, are not invisible after all.

Thus I have met, one after the other, Tom, Rankin, and Shellard. Hours have passed, and I still haven't talked to the professor, though I have seen him as my tour passed repeatedly by the open door of his office. A bit impatient at this point, I ask Shellard when I will be able to meet Professor Hawking and if he thinks it might be possible to do so now. He gets up quickly and apologizes. "Of course! It's more important to interview Stephen Hawking than me." I wait in his office. I'm wondering why Hawking has agreed to meet me. His personal assistant confirmed that it is quite rare for him to accept interviews. Did he hear about my study? Does he want me to stop investigating him? How am I going to behave before this person I have tried to demystify but who nevertheless is still very intimidating?[6] How are we going to communicate? Is he going to use his artificial voice? Shellard brings an end to all these thoughts: "Stephen Hawking is ready to meet you." I follow Shellard into the office. Hawking, sitting in his wheelchair, is half hidden behind a large computer placed on the desk in front of him. I think about what Gene Stone said: "Almost everyone's first contact with him is metal."[7] I, however, am struck and rather moved by his fragility. I greet the professor. Shellard, seeing that I don't know where to sit, asks Stephen and then tells me to sit near the professor. I ask if he would mind if I recorded our interview. Shellard seems to think that it's not a problem. With a smile on his face, he leaves the room. I can hear the nurse behind me as she comes and goes—washing, straightening papers, and so forth. I have decided to focus my interview on his role as Lucasian Professor and not ask him any personal questions related to my interest in his way of working. I switch on the tape recorder and open the conversation.

HM: Thank you very much for accepting this interview, Professor, I know that you are very busy. Tom explained to you why I wanted to talk to you.

I stop. I'm not sure that he is ready. He seems to write something, but I don't know where. After a few seconds of silence, I start again.

HM: Thank you very ...

His answer to my first question stops me. A voice speaks from somewhere:

SH: *I can give you a printout of what I say.*

HM: Thank you. It would be very helpful.

The tape recorder will nevertheless be on during the first part of the interview; I will be able to hear just the *click-click* of his commutator (the instrument he uses to spell out words) and my own questions.[8]

HM: So, I told you I've been commissioned to write an article in a book on the history of the Lucasian Professor of Mathematics, so my chapter will be on you, on Professor Hawking; the idea of the book will be to examine the role of the professor inside his cultural, intellectual, technological context. And so my first question is: Why is it so important for you to be the Lucasian Professor of Mathematics?

He begins writing. I realize that I force my voice to speak louder (strangely, I have the feeling that because he can't talk, he also can't hear). A sentence is written on the small computer attached to the arm of the chair (I'm not sure whether it was written on the big screen in front of him, because I didn't see it immediately). It is accompanied by the professor's synthetic voice. I will notice only later when I listen to the tape that his synthetic voice repeated the sentence too; at the time, for some reason, I didn't hear it.

SH: *Why you?*

HM: Why you? Ah! Why me?

I'm surprised; I was expecting the answer to my question and not a question from him.

HM: Because I'm a philosopher of science, I'm a sociologist of science, I work at the Maison Française of Oxford, I'm running a program in History of Science, and I'm extremely interested in creativity and discovery. So I'm writing a book.... I studied an inventor in France, and I was in-

> terested in your discoveries because you are considered to be someone very creative, and so I was told it would be appropriate if I could write this chapter.

As I speak he is still working at answering my question. I can watch the answer slowly appearing on the screen:

> SH: I didn't apply to be Lucasian Professor. I was already an ad hominem professor at Cambridge so it wouldn't have raised my salary.

The nurse asks me if I would like some tea. I say, "No, no thank you," and continue reading the answer:

> SH: I hoped the position would have been used to bring in someone good from outside Cambridge like atyia [*sic*].[9] I was rather disappointed when Georges batchelor, who was then head of DAMTP, told me the electors had chosen me. I urged them to think again and get someone like atiya. But Batchelor was very much against the idea of giving the Lucasian professor to Atiya, whom he regarded as a pure mathematician thinly disguised as a physicist. It might have meant that the Lucasian Professor was lost from applied mathematics to pure. In the end I agreed to accept if I was given certain support. I must say I like the feeling that I hold the same job as newton and Dirac.

For more than two hours, questions and answers follow, one after another. Five exactly. Five of the ten possible questions prepared in advance with the editor of the book. Five answers slowly written on the computer. At the end I leave with Hawking's answers printed out, with no trace of the "small talk" framing the interview (like Hawking's question "Why you?" at the beginning and his assertion, "My answers were hard hitting and frank—I don't want to water them down"). No trace of a whole range of interactions that occurred, through or thanks to the computer, which I will recall and describe here.

5:30 p.m. The room is now totally quiet. I'm alone with the professor. The nurse sits at the entrance of the room. She says hello to someone who just walked by the door, talks quickly with Tom, then concentrates on her knitting needle. Both Hawking and I sit in silence looking at the computer. I have just asked him if he thought that the intellectual authority of the Lucasian Professors of Mathematics had changed. He writes. I wait, following the cursor on the screen:

> SH: It is nice to feel that one holds the same position as newton and Dirac, but . . . [suction].

Suddenly, the *click-click* of the commutator stops. In the middle of writing his answer, this word crops up in his text, and the voice says, "*Suction*." The nurse immediately arrives, takes a Kleenex, and wipes his mouth. Then she opens a small bag, takes a little pipe, puts it in the professor's neck, and activates a machine. His body shakes. I think that I should perhaps leave the room, but the professor, seemingly oblivious to the nurse, continues with his writing:

> SH: . . . but the real challenge is to do work that is even a small fraction as significant.

His body has been taken care of and he can carry on with what he was doing: being the Lucasian Professor of Mathematics writing about the Lucasian Professor of Mathematics.

I ask him a new question: "How do you imagine the future of the Lucasian Professor of Mathematics?" Hawking starts writing.

> SH: I think I was appointed as a stopgap to fill the chair as someone whose work would not disgrace the standards expected of the Lucasian chair, but I think they thought I wouldn't live very long, and then they choose again, by which time there might be a more suitable candidate. Well I'm sorry to disappoint the electors. I have been Lucasian professor for 19 years and I have every intention of surviving another 11 to the retiring age. Even so, I won't match Dirac, who was Lucasian professor for 37 years or Stokes who was for 54. [legs]

Again, a word appears under the sentence he has just written. A voice says: "*Legs*." The nurse arrives, delicately takes his leg, and moves it slowly. Then "Chair-in" appears on the computer screen (though the voice stays silent). The nurse asks him if it concerns the batteries. She looks at the little computer resting on the arm of the wheelchair, then turns the wheelchair around and begins to manipulate a big black box under it marked MOZART. The voice says, "*No*," then, "*Get Tom*." The nurse goes out and comes back with Tom. She explains to him that Stephen must have a problem; she thought that it was the battery. Tom looks quickly at the computer and then inspects something under Hawking's chair. He leaves the room and comes back with a big black box with *Mahler* written on it. He replaces the black box settled under Hawking's chair with the new one. Hawking can now start writing again and finishes his response.

> SH: I hope they get someone good after me, but now that there are so many professors, it won't have quite the same draw.

I ask the professor if he feels too tired or if I can ask him another question. The voice says, "*Yes*." I ask him my question: "You talk a lot about Newton and Dirac. What do you think about Babbage?"

The thumb starts caressing the commutator in his hand again; the words accumulate one after another on the screen. After fifteen minutes, the cursor stops moving. I never know if he has finished or not. I can read the answer:

> SH: I'm also proud to have babagge [*sic*] as one of my predecessors, even if he wasn't Lucasian professor for very long. I do a certain amount of fund rising [*sic*] for Cambridge university in places like silicon valley, and [it] is a good line to throw in that the father of computers was one my forebears.

It's almost 7:30 p.m., so I ask him if he wants me to leave. He writes, "Yes, I have to go." Then, "My answers were hard hitting and frank—I don't want you to water them down." I ask him if he wants me to show him the article before its publication. The voice says, "*Yes*." He writes, "When will the paper be finished?" I say, "Not before the end of next year." He smiles. (I interpret this smile as saying, "You are as slow as I am"). I say, "Good-bye and thank you." He writes on the computer, "Can you get the nurse." He sees that I read the message and the voice is not activated. I go to get the nurse, but she's not there. I try not to get panicky and decide to get Tom in his office. I open his door; he's making a phone call; I tell him that Professor Hawking wants his nurse, but I can't find her. He immediately ends his call and goes to see Hawking. I follow him. When we arrive in the room the nurse is there. "So you were looking for me!" she says. She shows him some pictures, and he smiles. I seize the opportunity to look at the copy of the famous bet he made with Kip Thorne about one of his hypotheses, which is hanging on the wall.[10] The nurse notices that I'm still here and asks me if I want something. I say that I wanted to say good-bye and thank the professor. I leave and go to see Tom. He leaves me briefly to pick up the printed answers to my interview. When he comes back, I mention that the professor agreed to answer some of my questions by e-mail. Tom adds that if I need anything, I should write to him, and he will relay it to Hawking or organize interviews with other people I want to meet. It is time to leave. The cars I saw through the windows when I arrived three hours earlier have disappeared. Instead, I see our reflections in the windowpanes. The sun has set. I leave Tom and walk through the now silent and empty waiting room. I close the door of the Department of Applied Mathematics and Theoretical Physics, leaving behind me the well-protected building that houses the Lucasian Professor of Mathematics. The place is once again shrouded in mystery in the twilight of an English summer evening. That was June 25, 1998, at 8:00 p.m.

READING THE *MISE-EN-PRÉSENCE*: A BODY AND A TEXT

I now would like to explore the key role of the computer as the physical—and not metaphorical—extension of Hawking's body in the scene described above, especially with regard to the place it occupies. I am not face-to-face with Hawking, but, like him, seated facing his computer. Or, rather, my body is facing the computer, whereas my face is turned toward Hawking when I ask him questions (as I would in any interaction), for it is him and not the computer I am addressing. My head turns slightly, in line with my body, to read his answers displayed on the screen, that is to say, to read what I interpret as being his wishes, will, and thoughts. As for Hawking, he remains seated, an extension of the computer to which he is connected by his finger on the commutator. At times he tries, almost imperceptibly, to move his head, which is already leaning slightly toward me, to give a very faint smile as I talk to him. It is through this constant coming and going between a silent human body, capable of "talking" with its face (that is to say, of creating a presence, a contact), and a more-or-less silent mechanical body[11]—depending on whether or not Hawking decides, after having made his discourse visible, to make the machine "talk"—that I see the first link between these two bodies emerge. Without the computer I would understand nothing of what Hawking wants or wishes to express. Without Hawking, the computer would remain opaque and voiceless, and my presence would be incongruous.[12] Yet because of this triangulation (Hawking, the computer, and myself), the reading of the interaction—that is, of what is wanted, said, and expected by the different actors—is totally muddled. In the first minutes, I don't know whether I should look at Hawking, the computer in front of him, or the one attached to his wheelchair. I can't hear the voice talking to me; I don't know whether or not Hawking has finished writing, whether I can or should laugh, wait, or carry on asking questions, stay or leave.

The practices of attuning to a conversation that we take for granted when we talk to someone endowed with similar competencies are disrupted and put together again in different ways. Hawking's unique conditions, the distinctive modalities with which he communicates and my unfamiliarity with them, thus tells us something about what is implied in a "normal" interaction or conversation.[13] We find ourselves confronted with a paradox. On the one hand, we have shown that Hawking, because of his handicap, makes visible that which is normally hidden in the scientist's practice and self-presentation, for example, the role of assistants and machines to which he is attached.[14] On the other hand, his handicap also makes visible, because they are absent or because they are modified by his *mise-en-présence*, a set of elements that are visible (although unnoticed) constituents of all social interaction, such as body language, intonation, the passage of time, and spatial relations.[15] Let us start with time.

Conversation analysts have shown that "the organization of taking turns to talk is fundamental to conversation, as well as other speech-exchange systems."[16] The turns that an "interview system" organizes are formalized as "questions" and "answers." In this context, Hawking's response to my first question occurs ten minutes after I asked it—and after my second question—as I was repeating it for the second time. And, instead of responding to my third question, he asked me a question instead. The normal order of question and answer seems inverted.[17] It is known that "discontinuities occur when, at some transition-relevance place, a current speaker has stopped, no speaker starts (or continues), and the ensuing space of nontalk constitutes itself as more than a gap—not a gap, but a lapse."[18] Hawking's profound silence, when it occurs at transition-relevance places, becomes a lapse, which creates a discontinuity in the flow of the conversation.[19] Because of the organization of turn-taking, "a participant, willing to speak next if selected to do so, will need to listen to each utterance and analyse it at least to find whether or not it selects him as next speaker."[20] In this case, we are not in the relation of talk-listen-talk, but of talk-read-talk (plus sometimes "listen"), and one notices the oddity of reading and listening at the same time. Because the question "Why you?" written on his computer is accompanied not by his voice but by an artificial voice, and my eyes are not trained to read this interaction, I won't be able to know that it is addressed to me.

Moreover, one is struck (because of their absence) by the importance of the markers inscribed in the body, which enable us to open, sustain, and close the interaction.[21] For Merleau-Ponty, "The body appears as a phenomenon with which we are familiar; we perceive human behavior through the body.... The body, having a higher degree of integration by reason of its structuration as a *human* body, is the visible expression of meaningful behavior, communicable to other 'ego-bodies' which are, likewise, centers of meaning and points of mutual encounters. This already implies an intersubjectivity, ... but actually it will be seen that the encounter is necessary for the very 'realization' of the subject."[22] Put another way, Erving Goffman has shown that what constitutes and makes face-to-face interaction possible is precisely the immediacy of the corporeal co-presence, which implies using signs or clues displayed by bodies in order to regulate the interaction.[23] With Hawking, all the markers "normally" inscribed in a body (gestures, glances, positions, intonation, silences, sighs, or throat-clearing) are missing. Instead, I have a more or less approving smile on a statuary body and a discourse fixed on a machine, punctuated by signs relating either to physical needs (e.g., "legs," "suction"), to wishes, feelings, or thoughts (e.g., "I hope," "I think," "I am"), to questions (e.g., "Why you?"), or to orders ("Get Tom!"). Finally, an important element of an interaction (especially for turn-taking) is mutual perception, which is made impossible in this context because of the mediation of the computer. We don't look at each other; we look at the computer instead.[24]

If we follow Goffman, gestural movements are signifying movements; they have a conventional meaning. A smile does not have the same meaning in Japan as it does in America. But, more than that, gestures reveal a reality that becomes visible only through their mediation, which is that of a person. Bodies, in other words, become surfaces on which the definitions of the self and the other are inscribed. This piling up of mutual self-definitions (and denunciation or ratification of these definitions in the course of the interaction) takes place "between" bodies. My first observation, then, is that this human body, through the mediation of the other body (the computer) talks only about itself and tells me nothing about myself. That is why I don't know what to do. I'm interacting not with *a* body but with an *extended body*.[25]

In other words, I am confronted with the following:

1. *An "illegible" body*, for our eyes are not used to interpreting a body without movements, gestures, and a voice that betray something about their author and enable us to attribute intentions to it and to read in it whatever it incorporates and says about ourselves.
2. *A computer* that, by its presence, reconfigures the spatial arrangement necessary for the mutual perception that makes an interaction possible and that, by the waiting required by the task of writing, destroys the timescale necessary for the smooth running of a conversation. For example, answers do not directly follow the questions preceding them. Moreover, the artificial voice is "next to" Hawking, where it repeats a written discourse. By acting as the echo of the writing, the voice disturbs understanding. It repeats but does not talk. Finally, functions generally dissociated in a conversation, like listening and reading, are simultaneously combined.

3. *A discourse*, which is "difficult to read." According to Paul Ricoeur,

> Dialogue is an exchange of questions and answers; there is no exchange of this sort between the writer and the reader. The writer does not respond to the reader. Rather, the book divides the act of writing and the act of reading into two sides, between which there is no communication. The reader is absent from the act of writing; the writer is absent from the act of reading. The text thus produces a double eclipse of the reader and the writer. It thereby replaces the relation of dialogue, which directly connects the voice of one to the hearing of the other.[26]

Although written, this discourse does not have the same modalities as writing, such as a certain fluidity, insofar as it is detachable from its conditions of production, and the inclusion of an "invisible" and "unknown" reader; in this case, this text is intended for me; it "speaks" to me. Neither does it have the modalities of a conversation. In a spoken discourse, "the subjective intention of the speaking subject and the meaning of the discourse overlap each other in such way that it is the same thing to understand what the speaker means and what his discourse means. . . . It is almost the same thing to ask 'What do you mean?' and 'What does that mean?'"[27] But, in the case of Hawking, this equivalency is ruptured. The discourse does not come from the mouth of its author; his *I* is in front of him—on the screen—and the markers of turn-taking are absent. Moreover, what this writing does fix is not the "said" of speaking but the event of speaking.[28] Similarly, this discourse fails to show the modalities of interaction by, for example, e-mail, for the author is present—juxtaposed to his text—and at my side. I therefore have to read a text without having the punctuation of the author's voice, yet with the silent author beside me.

What is "normally" attributed to the body, the machine, or the mind is inverted. The specificity of a machine (as opposed to a body) is that it cannot repair itself; this is precisely what characterizes Hawking's body. The specificity of a mind is that it produces thoughts, language, desires; in Hawking's case they are legible and emanate from the machine. So where is his body?

LEARNING ABOUT HAWKING'S EXTENDED BODY IN SITU

Moments of crisis shed the first light on the matter. When something goes wrong with Hawking's body (pain, discomfort) or with the computer (failure), the assistants—*ever-present* and ready to intervene, although discreet and relatively unobtrusive during the interview—are mobilized by the computer and the activation of its voice function. They try to be immediately operational by reading the words displayed on the computer screen, or by reading Hawking's face, or by touching the various apparatuses surrounding him, just as a doctor would feel the various parts of a patient's body to detect painful areas. Hawking is at the same time the surgeon in command of the bodies of others and the patient being taken care of by the others.[29] Thus the parts of the human or mechanical body to be "repaired" are identified ("*Legs*" or "Chair-in") and the operations to be carried out are indicated ("*Suction*" or "*Get Tom*"). In this process Hawking's voice acts as an indicator: "*Yes*," "*No*," and so on.

The distribution of tasks seems fairly well codified, though it can also be negotiated during the action. When the voice says, "*Legs*" or "*Suction*," there is no ambiguity. The nurse is immediately operational, whereas "Chair-in" causes a moment's hesitation. The nurse wants to extend her competence to repair the computer but, as we have seen, Hawking does not want her to intervene. That is Tom's domain. The artificial voice is important because it is used to alert this small collective of assistants when there is a problem. It thus enables them to have a relatively large degree of autonomy. They are not forced to be constantly breathing down Hawking's neck. The nurse, alerted by the voice, can serve as a relay between Hawking and Tom. It is also through this voice that a semblance of "normal" conversation with me can be created ("*Yes*," "*No*," "*Why you?*"). He does not, however, use it to provide formal answers to my questions or to communicate with his assistants when they are around him. Thus, communication with Hawking involves the reading of signs on the computer and signs expressed in body language (the movement of his eyebrows or mouth)—skills acquired over time by the people working with him. Between body language and formal writing there is a gap: that of "speaking." We also notice a strange inversion: a deaf person would read the lips of the person who speaks; the assistant reads Hawking's face. As Tom put it:

> TA: He doesn't need to use his voice, really, unless I'm facing him and on the table, he talks with his face, when I am talking about work or whatever, or doing slides for him, *there's a lot of putting words in Stephen's mouth*, and he says that's a yes and that's a no, there's a lot of that; . . . so, yeah, he talks with his face.
>
> HM: So, I imagine that you anticipate a lot!
>
> TA: Yes, there aren't many answers. . . . Like if I say, "Do you want this published?" I'll see how he reacts. Or if he's talking about a meeting, and I go, "Tomorrow?" . . . It's pretty obvious he either wants it tomorrow or after he

> gets back.... You don't have to wait for him to say it, and if there's a time, I'll suggest a time, and I'll suggest other times.... But, yes, you put words in his mouth, and he communicates with his face a lot, so there's a lot of that; *he doesn't need to use his voice, as you look on the screen....* You will see that, surely.

As we have seen, Hawking's special conditions impose the mechanization (the hierarchization, standardization, and routinization) of his human/machine-based environment.[30] What is at stake behind this coordination and hierarchization of tasks is human life. A false maneuver in any of these operations can result in death.[31] The graduate assistant will characterize his job not only by its high level of responsibility (in the end what is at stake is the life of a human being) but also by the availability required ("We are on call twenty-four hours a day") and the difficulty of living constantly together. What is also at stake is Hawking's ability to work and to maintain his status as STEPHEN HAWKING the genius physicist.

HAWKING'S THREE BODIES

"To use language in speech, reading and writing, is to extend our bodily equipment and become intelligent human beings. We may say that when we learn to use language, or a probe, or a tool, and thus make ourselves aware of these things as we are of our body, we *interiorize* these things and *make ourselves dwell in them*."[32]

As we have seen, the computer has thus become an appendix of Hawking's body, owing to the acquisition of a number of skills that Hawking has mastered: the rapidity with which he touches his commutator, the manipulation of certain operations for selecting words in the construction of sentences, and the choice of the right keys to activate the synthetic voice. As one would say of an individual who has mastered all the subtleties of a new language, he has, as one of his students points out, "become totally 'fluent' with his computer." Remember that the designer of Hawking's software, Walter Woltosz, characterizes the computer as being "really an extension of himself now."[33]

According to Polanyi, "Every time we assimilate a tool to our body our identity undergoes some change; our person expands into new modes of being";[34] or, as Drew Leder suggests, in the tradition of Merleau-Ponty, "The true relation between body and instrument, phenomenological rather than crudely materialistic in character, is only revealed when we reverse the analogy. It is not that the body is like a tool, but that the tool is like a second sort of body, incorporated into and extending our corporeal powers."[35] "To incorporate a tool is to redesign one's extended body until its extremities expressly mesh with the world."[36] In this sense, this tool, the extension of his human body, enables Hawking to mobilize and extend his competencies by affording him the ability to select information, read and write scientific articles, prepare talks, and communicate with colleagues informally or by e-mail conferences. As he himself put it in a recent interview conducted via e-mail, "Without this software or something like it, I would have been cut off and unable to carry on as a physicist."[37]

However, our analogy with the world of phenomenology stops here. Indeed, for Polanyi, the body is, above all, *the* main instrument through which the world is apprehended: "Our body is always in use as the basic instrument of our intellectual and practical control over our surroundings."[38] "Our body is the ultimate instrument of all our external knowledge, whether intellectual or practical."[39] In this particular context, we have seen that it is not Hawking's flesh-and-blood body that becomes "the basic instrument of [his] intellectual and practical control over [his] surroundings" but the computer itself. The computer is the center of the network through which (almost) all interactions are negotiated, because they are rendered visible and public (the computer distributes competencies between what is appropriate to or for Hawking's flesh-and-blood body, his identity, his assistants, his computer, and me). The computer makes the interaction possible between Hawking and myself (Hawking can write, answer my questions, and create a semblance of conversation), between Hawking and his assistants (he can mobilize them, point out problems that interrupt the smooth functioning of this interaction), and between Hawking and himself (it allows him to write his flesh-and-blood body). By allowing the mobilization of assistants, it allows the maintenance of the flesh-and-blood body and of the machine, and, ultimately, the writing of a text.

Hawking can move only one finger and a few muscles of his face. The structure of the system (the way in which language appears and is organized in and on his computer, such as the alphabetical organization of words in columns, the gathering of a vocabulary most commonly used, possible abbreviations, word predictions, instant phrases, his artificial voice) and everything it enables him to do participate in the construction of his intellectual competencies as much as they extend them. In other words, these instruments are not merely transparent extensions through which his intelligence is diffused, for they comprise it as much as they allow it to be diffused.

The body is the privileged instrument or, as both Polanyi and Merleau-Ponty say, trying again to muddle the dichotomy between instruments and the body, instruments are also another part of the body. With Hawking, the body is not the privileged instrument, and the instruments are much more than another part of the body; they make the body present as much as they comprise it. As we have seen, this system enables him to mediate his relationship with his own human body by excorporating or making visible the normally invisible functioning or dysfunctioning of certain neuromotor operations. Owing to the computer's mediation, Hawking can immediately allow the pain in his human body to be located, by writing "Legs," for example, on his computer. By activating the synthetic voice, he can mobilize the competencies of his nurse so that she is immediately operational to remedy this dysfunctioning. The human body is thus taken care of by the nurse, through the computer, which serves to alert her and to pinpoint the problem. In this way, through these different mediations, the functioning of this human body becomes visible (for us as observers and for the nurse), externalized, and collectivized to such an extent that, as long as the body is taken care of by the relay of machines, men, and women, Hawking no longer thinks about it (as when we swallow without thinking about it).[40] This sequence consists of a multitude of actions: feeling of pain in the human body "Legs" displayed on the computer, "*Legs*" spoken by the computer, mobilization of the nurse, movement of Hawking's leg, activation of his nurse's arm, leg rested—all of which have become the movement of one and the same self-repairing body. For Hawking, the movement of this natural, collectivized body has become invisible, as a body endowed with its own ability to move would be.[41]

On the other hand, the computer also makes it possible to visualize all dysfunction in the mechanical body (the computer is reflexive or says something about itself), for example, "Chair-in." Yet we have seen that dysfunction in Hawking's human body does not preclude the activity of writing, whereas dysfunction in the mechanical body precludes all interaction. It therefore also precludes the possibility of mobilizing the assistants and allowing the human body or the mechanical body to be repaired, as well as all possibility of writing.[42] As I observe Hawking orchestrating the functioning of his extended body through the mediation of the computer, I cannot help but think of the materialization of the "pilot in his vessel," a metaphor used by Descartes to describe the difficult union of body and soul:

> Nature likewise teaches me by these sensations of pain, hunger, thirst, etc., that I am not only lodged in my body as a pilot in a vessel, but that I am besides so intimately conjoined, and as it were intermixed with it, that my mind and body compose a certain unity. For if this were not the case, I should not feel pain when my body is hurt, seeing I am merely a thinking thing, but should perceive the wound by the understanding alone, just

> as a pilot perceives by sight when any part of his vessel is damaged; and when my body has need of food or drink, I should have a clear knowledge of this, and not be made aware of it by the confused sensations of hunger and thirst: for, in truth, all these sensations of hunger, thirst, pain, etc., are nothing more than certain confused modes of thinking, arising from the union and apparent fusion of mind and body.[43]

Whereas Hawking seems to feel pain in his human body, the nurse and Tom (indispensable elements of his extended body) give us the impression that they perceive as he does, through the intermediary of the computer (another part of the extended body), "the wound by the understanding alone, just as a pilot perceives by sight when any part of his vessel is damaged."

By unfolding itself and making itself visible through the machine, Hawking's flesh-and-blood body calls into question that which constitutes the primacy of the body itself, distinguishing it definitively from other objects. The body can't be "a mere object of which he is conscious as he is of every other 'thing' in the world." If so, he would have "embraced the position of the 'pure subject' who is a *spectator* before the world and also before his own body which he considers as part of the world of objects,"[44] because "in so far as it sees or touches the world, my body can therefore be neither seen nor touched. What prevents its ever being an object, ever being 'completely constituted' is that it is that by which there are objects. It is neither tangible nor visible in so far as it is that which sees and touches."[45] Indeed, "I am not *in front of* my body, I am in it, or rather I am it."[46] In other words, the perceiving body is always behind the perceived object, which is why my body "offers itself obstinately 'on the same side' without my being able to go around it."[47]

In a way, we can say that Hawking's flesh-and-blood body is registering in the same text as his speech; they both appear on the screen in front of him. Without this work of objectification of the body, Hawking cannot be his body. The body is felt but also represented as an exterior thing before it can do anything. With him, the distinction between body-object and body-subject disappears. The human body conceived as a body-object—that is, something that "allows between its parts or between itself and other objects only external or mechanical relations"—has become his own body, or the body-subject, that is, "'his' body as the center of 'his' existence, as the power both to act and to perceive and as a means for the subject to be part of the world."[48]

We thus see the movement of a certain form of absence-presence that characterizes the body as Drew Leder described it, but in another form. For

Leder, the body is always present but situated in the background. First, this is because our insides and even certain visible parts of our exterior, such as the area around our eyes, are hidden. Second, the possibility of allowing ourselves to absorb a situation is related to a form of forgetting ourselves. In situations of pain and discomfort, the body becomes present again. It is subjected to a form of "dys-appearance" or absence of absence. With Hawking, in moments of discomfort, the body becomes present again, but, more than that, it sets itself before him and before those taking care of "him" and it. It is thus collectivized, taken charge of, while the collective of humans and nonhumans becomes invisible again and disappears "behind him"; it has become his body. All his activity is then directed toward the activity of writing a text.

By incorporating the computer as the physical extension of his own body, Hawking can mobilize and extend his intellectual capabilities. He can "excorporate" and make visible the neuromotor functions of his human body,[49] as well as all the dysfunctions in the technical system surrounding him, and can mobilize the human competencies required for the smooth functioning of this collective. Finally, and consequently, he can simultaneously fulfill his role as Lucasian Professor of Mathematics. That is to say, he can write his own history by answering my questions on his role in the scientific community, the intellectual authority of his status, the changes it underwent over time, the characteristics he would like to see in his successors, and the importance of his peers' role as regards the authority of his own status. We thus witness the operation, maintenance, and *(non)simultaneous* articulation, through the mediation of the computer, of what I call, to paraphrase Ernst Kantorowicz, Hawking's three bodies: first, a *natural human body* collectivized—this natural human body cannot function without this collective; second, a *collective body* (human body/computer/assistants) naturalized—this collective body becomes "his" body; and, third, a *sacred body*—the representative of the Lucasian Chair of Mathematics (which "is not subject to Passions as the other is, nor to Death, for as to this Body the [Professor] never dies)."[50]

However, all the interactions described above relating to the functioning of this extended body are totally absent from the official text handed to me. Even though, as I tried to show previously, Hawking attaches his singular and mortal being to what we might call the Corporation of Scientific Geniuses by using his mortal body to place himself in the tradition of his peers, to establish the paternity of his ideas, or to justify the choice of his fields of research, he also effaces all physical evidence of his mortal body, despite its indispensability in the production of this paper. Indeed, Hawking is able to switch between the *talk* and *write* functions of the software. When the program is set to *write*, what he composes is stored in a file as in a word processor, but when set on *talk*, things are just written, said, and then cleared.

Thus, sleeping like the unconscious, the computer's memory is an immense repository of formal written articles, lectures, and interviews, while all the words concerning "informal" interactions evaporate, depending on whether or not Hawking deems them worthy of being saved. Accordingly, owing to the computer, I could "see" the (non)simultaneous functioning of these three bodies; and owing to the same computer (assuming, for example, that Hawking had sent me his history by post), we might not have been able to "see" how the dissociation between these three bodies takes place. That is, we would not see how Hawking, through the mediation of the machine, could both make his extended body function and, as he responded to my questions, construct his identity as STEPHEN HAWKING, Lucasian Professor of Mathematics, while at the same time, as the interview moved toward its completion, effacing all presence of his physical body, the people who surround him, and even myself. In his case, the "noise" and the conditions that made the interview possible were filtered out, not (just) afterward by the analyst, as is usually done, but during the interview itself. By the end of the interview, Hawking's responses—which appeared to me on his screen—were printed out as *sui generis* statements without any trace of their original conditions of production, while on my tape recorder only the sound of my own voice remained—questions asked to the invisible and echoed by silence. To paraphrase Barthes, what this machine of genius was supposed to produce was not only equations but also a (self-effacing) man (see fig. 6).[51]

EPILOGUE

We have thus followed a strange movement, a sort of coming and going between perceiving subjects and perceived objects, a zoom effect like that of a camera trying to set the right distance from its object in order to achieve a maximum degree of clarity—total transparency. On the one hand, we have seen how the tricky adjustment between the multiplicity of stabilized representations of Hawking and my face-to-face interaction with him has fragmented his identity. Like an impressionist painting, distance allows us to see what proximity causes us to lose. The layering of the multiplicity of narratives "about" Hawking reveals the stable identity of an individual that face-to-face contact does not allow. The closer we get to the object-subject, the further we move from it. We no longer know "who" or "where" he is. All the categories we normally use in thinking about a person, a body, a

I didn't apply to be Lucasian professor. I was already an ad hominen
professor at Cambridge so it wouldn't have raised my salary. I hoped
the position would have been used to bring in someone good from
outside Cambridge like atiyah. I was rather disappointed when George
batchelor who was then head of DAMTP told me the electors had chosen
me. I urged them to think again and get someone like atiyah. But
Batchelor was very much against the idea of giving the Lucasian
professor to Atiyah whom he regarded as a pure mathematician thinly
disguised as a physicist. It might have meant that the Lucasian
professor was lost from applied mathematics to pure. In the end I
agreed to accept if I was given certain support. I must say I like
the feeling that I hold the same job as newton and Dirac 11When I
joined DAMTP in 1962, it contained only two professors, the Lucasian
professor, Paul Dirac, and the Plumian professor, Fred Hoyle. Shortly
after that, there was a contest between Hoyle and Batchelor over who
should be head of department. Hoyle lost and in disgust transferred
his chair to the faculty of physics and chemistry. So for a time the
Lucasian professor was the only professor in DAMTP. But now we have
about five other established chairs and about six ad hominum
professors, the Lucasian professor ship is not so important. It is a
bit like the title, astronomer royal, that my colleague and fellow
research student, Martin Rees has 11It is nice to feel that one holds
the same position as newton and Dirac, but the real challenge is to
do work that is even a small fraction as significant.11I think I was
appointed as a stop gap to fill the chair as someone whose work would
not disgrace the standards expected of the Lucasian chair but I think
they thought I wouldn't live very long and then they choose again by
which time there might be a more suitable candidate. Well I'm sorry
to disappoint the electors. I have been Lucasian professor for 19
years and I have every intention of surviving another 11 to the
retiring age. Even so I won't match Dirac who was Lucasian professor
for 37 years or stokes who was for 54. I hope they get someone good
after me but now that there are so many other professors, it won't
have quite the same draw 11I'm also proud to have babbage as one of
my predecessors, even if he wasn't Lucasian professor for very long.
I do a certain amount of fund raising for Cambridge university in
places like silicon valley, and is a good line to throw in that the
father of computers was one of my forebears.

Figure 6. The printout of Hawking's responses to my questions.

machine, a mind, an interaction, a conversation, a text, and speech are blurred.

When we are in his presence, Hawking seems even more difficult to seize than we had imagined. The reason is that, first, as we have shown, the depth of the "social" body (the one that talks of the other as much as of itself and allows the accumulation of self-definitions), normally "visible" but "unnoticed" in any interaction, is "absent" in his case. Second, the internal functioning of the flesh-and-blood body, normally "invisible," becomes "visible" and "distributed" here (the "I" and "Legs" are mixed in the same discourse and are detached from the body; they are before him, while his voice is "next to him," and the flesh-and-blood body is collectively taken care of) and "omnipresent" (it appears at different moments during the interaction). Third, the machine, by the coordination of the actions it allows, disturbs the interaction as much as it creates it. This is where the confusion and fragmentation of identity is pursued by recomposition. Here we follow, step by step, how the machine to which this silent body is attached makes the components of a collective body visible and mobilizes a part of it—the assistants—to repair both the flesh-and-blood body and the mechanical body. We now see how this collective body becomes the man's own body (his extended body), while at the same time making possible the construction of his sacred—corporate—body as Lucasian Professor. Last, we are able to follow how the machine—another part of this collective body—gradually erases the presence of Hawking's flesh-and-blood body, his extended body, and my own body, all of which are conditions, in this interaction, for the construction of his sacred body. After having made visible what is normally invisible in an interview (the internal body), the machine makes invisible what is normally visible in the after-interview: its conditions of production. Thus, on the one hand, the machine disturbs the interaction as much as it creates it, while, on the other, the machine creates the interaction as much as it disassembles it, leaving that which will become public—a text that will be handed to me and stored in a part of the computer, waiting to be added to other narratives about Hawking and effacing that which has become private, the context of interaction.[52] Has Hawking eluded us again?

VI At the Beginning of Forever

Archiving HAWKING

Throughout this book we have seen the deployment of different parts of what I have called an extended body: the assistants and the machines (chapter 1), the students (chapter 2), the diagrams (chapter 3), the journalists (chapter 4), and the ethnographer (chapter 5). We could even talk of a multiplicity of extended bodies, each of which can be glimpsed through its workings and through its arrangements: bodies sometimes connected and sometimes divergent; bodies that become apparent or disappear, that exemplify or differentiate, that anchor or abstract.[1] The present chapter follows a parallel trajectory; in it we visit another world—that of the archivists, where we will similarly see a collective at work.[2] This Hawking-collective participates in the construction and "*mise-en-mémoire*" of HAWKING in the form of the Permanent Hawking Archive. But what is to be conserved? Can one collect "everything" that has been produced by and around him? What is important? And what is not? What is to be done with the multiplicity of similar articles written about and by him? Does the fact that everything is on his computer already make the necessity of an archive redundant? Or, put another way, how can archives be built when almost everything is already stored or "archived" in a computer? Conversely, what does it mean to archive electronic resources? Indeed, what here constitutes an archive? To answer these questions I follow the circulation of artifacts, texts, and images—both digitally and in print—that we have seen produced in the chapters above. I describe how these materials are interpreted, put into context, recorded, and stored for the sake of future generations and future interpretations. This raises a number of new and very specific questions that I address in this chapter. For instance, what happens to this material as it moves and circulates from the hands of the assistants to those of the archivists and from the hands of the archivists to those of the scholars, and then on to "the public" in general? What happens to this material as it moves from one medium to another? And what happens to it as it moves from the physical sciences to the archival sciences and from the archival sciences to the human and social sciences? In brief, how is Hawking's archival body constituted and preserved?[3] How is the present-absent author made? How is the self extended through time and space? How is Hawking going to be reconstructed through the texts he leaves behind?

THE BETTY AND GORDON MOORE LIBRARY

In 2001, thanks to a donation from Gordon and Betty Moore, Cambridge University was able to complete the construction of a new £7.5 million library devoted to the physical sciences, technology, and mathematics located adjacent to the Centre for Mathematical Sciences. Gordon Moore is credited with discovering "Moore's Law," and with cofounding Intel in 1968. The former predicted the exponential growth of computing power; the latter made this prediction a reality by introducing the microprocessor responsible for the information revolution in 1971. It is thanks to Dr. Moore that Hawking was endowed with a new voice: "I'm Intel inside," he likes to recall. This new voice enabled him to donate his "old" voice to the Science Museum in London. Similarly, Dr. Moore has made it possible for him to leave his papers to a permanent exhibition named the Hawking Archive at the Betty and Gordon Moore Library (the Moore). Situated on the first floor, facing the Isaac Newton Institute building, the Hawking Archive has been a key element in the original plans for the Moore since its inception. The archive is meant to instantiate not only a particular notion of progress, in which Hawking stands in the long line of the scientific greats, but as an emblem of the revolution in computer science that Moore inaugurated. As the head of the Moore told me, "Gordon Moore was looking for a fairly large donation, something which would survive as a real legacy for Cambridge, and so the idea of the Library appealed to him. [Given his] relationship with Stephen Hawking, . . . he very much liked the idea of bringing Stephen's material into the same environment as the previous holders of the professorship, Newton, etc. So the fact that the Library was part of the University Library fitted quite nicely into that."[4]

That the professor's archive will be housed at the Moore Library rather than the University Library is quite unusual. For some, the special circumstances in this case not only have to do with the donor of the library building,[5] but also with the fact that Hawking is such a well-known figure. Moreover, there are practical considerations. The University Library has never had a permanent display, and the Moore is much better equipped to host one. According to the Web site publicizing its construction, "The new Library will provide a home for Professor Stephen Hawking's papers and electronic archive, which he has offered to donate to the University. Initially, these papers will include Professor Hawking's hand-written material dating from before 1973 and an early draft of *A Brief History of Time*. In the longer term, the Library will provide a digitized archive for more recent material stored in electronic media by Professor Hawking. The archive will continue a great Cambridge tradition of preserving the papers of famous scientists for future study."[6]

Anthropologists have opened the doors to the multiple sites and practices that comprise the modern world in, for example, science, industry, administration, and finance (to cite but a few), but no anthropologist has produced a close study of the archivists' work.[7] In a certain sense, this work—that is, what the archivist does—provides the datum for narratives that have yet to be written; on the other hand, their practice is understood and motivated by Enlightenment narratives of scientific progress and perfectibility. Moreover, though much has been said about writing ethnography and writing history—or on collective memory—we know relatively little about the composition of archives in the "making" of history.[8] Let's then open the door, not of the laboratory this time, but of the archive.

The Moore Library's Head

The history of the "*mise-en-archive*" of Hawking's material coincides with the professor's move from his old office at Silver Street to his new office at Wilberforce Road. During the process, only the essentials were taken, leaving a great deal of material behind. His office—much larger than those of his colleagues—was emptied of all the material that had piled up on his shelves over more than twenty years—papers, books, and objects. This was boxed up, delivered to a special room, and stored at the Moore. As the head of the Moore Library told me, with a laugh, "So the opportunity turns up, the rest of the office turns up."[9] Later the curator of scientific manuscripts added that "it felt appropriate at the time, that the material that Stephen Hawking had collected in his office—or everything—all this material should be deposited as an archive, what *he* called an archive . . . or which was referred to as an archive."[10] More boxes from Hawking's new office, but also from his residence, continue to arrive at the Moore.[11] And "there *are* a lot of boxes!"[12] So far, everything is accepted. All this material is carefully documented as to its provenance. If an ex-nurse, for example, decides to bring certain items because Hawking didn't want them anymore, it will be noted that they arrived via this route and through this intermediary.

The senior librarian of science at Cambridge University and the head of the Moore and the Central Science Library is responsible for actively making contact with Hawking's entourage so as to receive materials. But, he says, "apart from being the main contact person, I don't have really any say over the long term of . . . the archive. It's been a responsibility split between the head of manuscripts [at the UL] and myself. It's just that I have been quite happy to be the splitting partner until I'm ready to release the material."[13]

> HM: Are they going to allow you to keep certain things?
>
> M: We negotiate. The head of manuscripts and I are kind of the joint keepers of it. So we have the original meeting, and, for the moment, he's quite happy not to have too much responsibility for it; I think he gets a bit more space [laugh].[14]

In the first stage, the material will be left in boxes. The head of the library hopes, as he says jokingly, that "it's a safe environment, [that it's] not [going] to be attacked by any bugs and so on."[15] In the second stage, he will be in charge of putting the material that has to be kept under proper archival conditions in the hands of the archivists at the University Library. This means that the material will be sorted out, catalogued, and stored on a particular type of shelving, in the right environment (i.e., no light, low temperature, and regulated humidity); that it will be protected against fire and thieves; and that, when permission is granted to access the material, it will be read under the supervision of a qualified and well-trained staff. We have entered here a sacred place, where what Derrida calls 'archontic' power—the control over authorship, accessibility, and interpretation—is exercised. We are also dealing with a kind of "paper organism" that needs to be preserved—it is (quite literally part of) Hawking's sacred body.[16] And just as the flesh-and-blood body is subject to the ravages of time and decay, so too are books and papers. Indeed, even those items seemingly immune to insects, worms, and the acidity of paper, like DVDs and hard drives, age and decay. In the third stage, it will be decided what can be left at the Moore, what "originals" at the University Library can be copied and displayed, and what can—or ought to—be digitized.

But before the material is released, the head of the Moore Library has to make it usable for the archivists at the University Library; that is, he has to make an inventory. His first reaction was that the amount of material *about* Hawking was proportionally much more voluminous than the material "created by him." Indeed, there was so little from him that when the head of manuscripts at the University Library came to have a quick look at what was there, he declared that there was not enough to constitute an archive: "Hawking doesn't really have an archive in any meaningful sense. Everything is done electronically, and I don't know to what extent that material is archived or in a state where it could be preserved permanently. . . . The strictly archival material is very, very little."[17] The books and articles *about* Hawking are thus not what this archivist would include in the archive; instead, for him, this is a library collection.[18] For him, what should be in the archive are Hawking's personal papers: his correspondence, manuscripts, typescripts, drafts, proofs of publications, and writings such as lectures, including material that might not have been published. For the head of the Moore, "It is the stuff generated by him, grant proposals . . . scientific papers . . . papers 'read by' . . . *that* material is quite a lot of material, by him, originating from him, [or that] he's responding to."[19] For the curator of scientific manuscripts, what would be of interest are the personally created items, though he will also take into account the library collection: "Well, you see, I mean, you could say that everything in that storeroom is essentially the archive that he kept, . . . the collection of his personal work. But, we're drawing another distinction or two. You have different skills to handle different things, and the individual unique items need to be handled in a different way than the material that's been published."[20]

At first, the head of the Moore and his assistants will sort through what is delivered to find out what's in this "mixed bags of things," such as books and articles about Hawking; cuttings (clippings) from national and international newspapers; grant applications, personal correspondence, fan letters, a few letters written by hand; objects like photographs, honorary degrees, medals, certificates, university documents offering him the professorship and the fellowship at Caius, a passport, a few Christmas cards, a commemorative folder of a trip, gowns, and so forth.[21]

Different skills have to be applied depending on whether they are mobilized to preserve books and published material, personal collections, manuscripts, and so on, or electronic material. Pictures and photographs present their own problems. Floppy disks, magnetic media, and tapes offer other difficulties. As they are obliged to look after them, they ought to regenerate them every five years or so; otherwise, they are likely to decay. At present, they are not equipped to do this. The same applies to movies, which have now been transferred to video, though not yet to DVD (which, in turn, will only last ten years). As the curator of scientific manuscripts says: "Most of the material that I deal with on a day-to-day basis doesn't present those problems. It's paper."[22]

Different skills mean that the material might be allotted to different departments. Keeping this in mind, archivists will classify the material at hand: on one side, the electronic material (AVIs, GPGs, PDFs, and document files, floppy disks, magnetic media, videotapes, and DVDs, etc.); on the other, printed material from and about Hawking (books, articles, magazines, newspapers, and newspaper cuttings), many of which also exist in electronic form; and lastly the nonwritten materials, that is, what they call artifacts: medals, gowns, and perhaps even wheelchairs and speech synthesizers. One can also find different drafts of work in progress in the form of computer printouts, preprints, speeches, and correspondence. Selection will occur here as well.

Magazine articles and books written *about* Hawking will be collected and assembled as part of the collection at the Moore. Published material, such as books *by* Hawking and different editions of his work, will be compiled into a bibliography and kept physically at the Moore with the class mark "Hawking collection." Foreign editions of his books are particularly interesting for the head of the library, because there is no other library that has a large collection of translations of the same book:

> So . . . that's a kind of significant end of the collection, and it just is a question of . . . do we get them all catalogued, or catalogue one copy, as part of the Hawking class marker University Library, for the sake of completeness. Yes, if you're interested in how the reception of this book was . . . actually, somebody donated last week a Thai translation of *A Brief History of Time* we didn't have. . . . So if somebody wants to study translations and how things have been, then, they are all in one place.[23]

The status that should be granted to the books that came *from* Hawking's office is still at stake. A lot of these books were sent to him, put on shelves, and probably not read by him. He may have autographed them, but they may have just been marked as his property later on by somebody else. How can one interpret the books that were in his possession? For the head of the Moore, it is "simple"; the selection is to be made by asking Hawking himself:

> There are books which are obviously key texts for when he was a student, but there are also books which have been sent and just stepped off the shelf, . . . so speculative people sending him a book, it's got a wrap or it's still in plastic or it's a review copy he's never reviewed. And it's a combination of all this material, which was in his office. At this stage we need to go back and say, "Now we have these things, are they *truly your* books or is this

> essentially junk mail ... [that we don't need to] save.' Everything which was in his Silver Street office has to be given a [trial] status.[24]

For the head of manuscripts, this is not an issue. Printed books just are not part of what he calls archival material:

> A lot of people sent him books. They sent copies of their works to him, and ... they [at the Moore] have a very large collection of those, but of course most of these books he probably never read, had no interests in (laugh), and they had no influence on him, so *it's a rather loose connection* with him.... *It's not an archive*, there is nothing to come here, they all remain in the Moore, these books, as far as I'm aware, and if they were ... they did come here, they'll go to the rare books department; *they wouldn't come to my department.*[25]

And he adds, "Well, the books we have from Darwin's library here are all annotated by Darwin. All of them.... We don't have the ones which don't have any annotations."[26] It is Darwin's annotations that transform books into manuscripts considered to be a legitimate part of the archive. However, we don't always write when we read, and Hawking couldn't annotate his books even if he wanted to. The only way one can know if these books were important for him is either by asking him or by asking those who select and interpret his work:

> All we know is that his wife told me that these books could not be regarded as representative of, you know, his views, or as having influenced his favorite things. They were mostly things that had been sent to him which he didn't necessarily agree with or have any interest in.... So she was rather dismissive of this ... group of material.... But, in any case ... that's an irrelevance from my point of view.... Printed books are ... not archival materials, and they don't form part of any Hawking archive ... held here.[27]

For the curator of scientific manuscripts, the fact that Hawking opened these books or not, read them or not, or was influenced by them or not, doesn't matter. What matters is that they were in his office—they shared the same physical space, and nobody took the initiative to discard them. Not doing anything is doing something. As far as he is concerned, these books could be kept at the University Library and catalogued by their own cataloguers and identified as a "special book collection," though they probably will stay at the Moore Library.

> I think its value, in library terms, is that it was Stephen Hawking's collection of what *he kept* on *his* shelves.... There must be quite a number of things that were sent to him ... that were just unpacked and put on the shelf.... He actually had it. That's all you can say.... It is established that he received it, and *he, they* didn't throw it away, or *they* didn't give it to a library or something.... You can't put too much interpretation into this, so most of what I'm saying is about the physical nature of this, or personal action, in some way, isn't it?[28]

Because the extended body—the personal assistants or the graduate assistants—didn't sort out or remove the material, it was interpreted as being "intentionally" kept and desired. Thus, should the archivists just establish a list of books that were in Stephen Hawking's office with a class-mark at the University Library under special collections? Do they actually need to keep the physical volumes? If they do, where should they be kept? The decision about what to do hasn't been made yet. In the meantime, the printed material will be put into proper archive boxes in a nice, secure, dark storeroom and kept at the Moore until the verdict is pronounced. The twenty or thirty boxes containing grant proposals will go to the University Library, as they are part of Hawking's personal creation. Finances are important resources for historians. They will be of potential interest for those who want to understand how Hawking's support network was maintained. In this regard, if the anthropologist is able to follow certain processes that will become invisible to the eyes of the historian—for example, how Hawking's documents are processed (and in part were produced)—the historian will be able to see what is invisible to the eyes of the ethnographer: grant proposals and the money that derived from them. As the head of the Moore says, "[The University Library] is used to dealing with that stuff. They know exactly what to do with it."[29] It will be a similar process for personal papers and the general correspondence (about ten boxes): "They can probably archive them. We are doing a kind of rough indexing job first, just to make sure what we got.... They will look after them."[30] The most recent files of general correspondence are in discrete archival boxes (discussion papers, personal arrangements for trips and visits, book tours, interviews, and correspondence about making films about "the book").

Though they are not considered part of what constitutes the archive "of a scientist," a sample of his fan letters demonstrating "the popular" side of his work will be displayed at the Moore. It seems that HAWKING, the celebrated genius, and Hawking, the working scientist, are being made into separate and autonomous entities, thus paralleling another distinction embraced by the archivists, that between "science" and "its context." The anthropologist of science sees here the construction not simply of an archive but of science in the making. As the head of the manuscript department told me, "I don't think members of the general public writing in to him, that's really not very interesting, every well-known person must get a lot of letters of this type, really these letters have nothing to do with *him*, *he's just receiving them because he is well known*, but I think in the case of an important scientist, obviously, correspondence with other scientists is important."[31] Thus, they will be either discarded or sent to the University Development Office, which might use them for possible fund-raising prospects:

> Yes, we got a sample of just the unusual daily post of the strange material which turns up on a daily basis. And there are a lot of samples of that. Which is basically a couple of days intake, two huge envelopes full of very strange letters. . . . At the very early stage we were [asked] by his personal assistants, "Do you really want to get all of these? . . . *Because it's going to be a bag per week* or something like that." So the librarian said, "No, not really, but we might keep a selection just to show this is the type of thing he received."[32]

Accordingly, they also will go through the letters and general correspondence and decide what they will want to display in the Hawking Archive. They will then negotiate with the University Library to send them a replica or a copy of the original material that is stored there under proper archival conditions: "And the early papers, it's just going through looking for interesting things which we might like to feature in an exhibition here, some of the originals, we [will] put a replica on display probably, because the building is twenty-four-hour access, and so security for anything more precious is a bit . . ."[33]

The nonwritten material, that is, the artifacts (medals, the statue of Hawking as he appeared in the Simpson, etc.) will be kept at the Moore. They are "visually interesting" and could be displayed in the exhibition along with photographs and copies of Hawking's publications. The "originals," his handwritten papers, for example, will be copied and displayed as artifacts or relics of a body that wrote by hand and vanished long ago.

In the decision-making process, the head of the Moore Library will be the principal interlocutor for Hawking and his family, who will have the last say. The selection of elements (manuscript and papers) that make Hawking HAWKING is based on what we already know about him and upon how he (and his entourage) want him to be remembered. Hawking's wife, for example, will want to emphasize the popular aspects of his work, while Hawking will want to enshrine his status as a scientist. One could argue, paraphrasing Michael Lynch, that the archive embodies an intentional design: it is "tightly constructed to enhance the reputation of an author."[34] As the head of the Moore said,

> I had a meeting with Mrs. Hawking, and she had ideas about what the current display should be like, and, I mean, . . . the popular side, but he [Hawking] also likes to kind of display what are his most important works that he'd like to be remembered by. . . . It's finding a balance where it's gonna get some recognition for people as they pass through, but also it stresses what his work was, the fact the he wasn't just a popularizer, and just didn't want to be remembered for one book in the end.[35]

Hawking once came in person to express what he wanted to be remembered for. He was presented with a list of publications and selected those that he thought would be the most relevant and influential: for example, the typescript of *A Brief History of Time* (the only item currently at the University Library) and some of his early papers, such as his paper on Hawking radiation.

> M: Yes, we did some change on display and put it up there, so put the copy of the article . . . the famous papers . . . (laugh . . .)
>
> HM: Is it there actually?
>
> M: Umm. I think I put one up.[36]

There are clear advantages to the fact that Hawking is still alive (most individual archives are constructed after their progenitors have died): "It's a lot easier for us to deal with Professor Hawking and his wife who are, . . . I mean, they still have material at home, and they are going through some of that and bringing it in," as the head of the Moore said.[37] It can also be helpful in the general process of the *mise-en-contexte* of these papers and objects. Part of the job of Hawking's graduate assistant, on top of all his other duties, is to help the professor intervene in this collective process of memorializing his identity. The head of the Moore tells me that Hawking is able to provide the context for some of the materials still in his house, "because obviously he is the one who knows about it. Otherwise all these photographs become very mysterious. Is it a conference? Is it a tour? Is it in Korea? Is it in Japan? What year is this? So the identifying of similar material is fairly important, . . . but also it's very tiring work."[38]

Hawking was also helpful in making sure that Gordon Moore would donate the money to create the library that houses his work: "Gordon Moore liked the idea of there being a named archive, . . . so he was keen in that there would be some permanent exhibition of some sort. That was useful. And *Stephen was the implement, watching on and making sure that we could get the money from Gordon Moore*, so that was good that the center as a whole all came together so that we were in position to collect the material when the building was finished. We will be storing the archive material in the science library rather than several places."[39]

However, if dealing with someone who is still alive can sometimes facilitate the task, it can also disrupt the smooth process of circulation of papers and objects. Indeed, Hawking often resists his *mise-en-archive*:

> It's more delicate when you're dealing with the archive with people who are still alive. Because *they give things and then they want something back*. Or they want something in the office, because we have them. Oh, he has a blackboard, he donated the chalk blackboard for the conference, and it . . . hadn't been treated [right]; it was just chalk from the conference which was . . . like twenty-five years ago. But it had gone from his old office to his new office, and from his new office to here, upstairs and into the storeroom, and then back down again across the courtyard, and back to his office again. So . . . I guess, it's a delicate kind of balance, but, obviously, . . . he wanted it back, so it went back. And it is on the wall. And he likes it for sentimental reasons. And I think this library, we would probably be stuck with that, because that's a very unusual archiving object [laughter].[40]

In effect, so long as Hawking is alive, the archivists are not dealing with "an archive" in the strict sense of the term. The "archival material" will be incomplete until Hawking dies and all the materials presently in his possession are inventoried and stored away, thus "completing" the picture, and constituting a "true" archive.[41] One would imagine that archives also continue expanding well after the death of the author, as new caches of letters or other documents are found, collated, organized, and added to the archive. As the head of the Moore Library told me, "There might be some early [material about his computer] that would turn up as we went through the correspondence. Because obviously, *it's partial* because he will have some material where he wants to keep some correspondence—certain correspondence—going back all the way, or he's got whatever his current files are of the personal ones at home."[42]

Though unusual, part of the material will be left at the Moore thanks to its founder, who not only supplied the computer technology enabling Hawking to communicate, and therefore create his own archive, but also the physical space within which to house it. Indeed, if an e-mail has been printed out and filed with the other archival material in the Moore, the University Library will take that. If the e-mail is sitting on the computers in Hawking's office, they obviously can't take it. Eventually, this will be part of the electronic archive. Nobody knows at this point how much of his e-mail correspondence survives.

Finally, Hawking will decide when and how the material will be available: "The other thing is, at that stage we have to sort out what terms of access [are] acceptable with this material, . . . but obviously you have got a subject who is still alive, who has some rights to say how accessible the material is going to be [and] when it could be accessible or not."[43]

Concerning the general correspondence, the main organizational criterion is chronology. That's the main job of the head librarian's assistant. The tasks are relatively well defined between the University Library, the head of the Moore, and the assistant librarian. The archivists at the University Library will be receiving Hawking's material from the Moore after it has been identified (or not) for preservation. The head of the Moore will be the relay between Hawking's family, his assistant, and the University Library. The assistant, under his supervision, will be in charge of identifying, classifying, organizing, and labeling what each box contains. As for Hawking, we have seen that his presence makes a difference in the actual localization, circulation, and redistribution of what will be stored or displayed. He will have a partial say as to what should be displayed in the exhibition and how the material should be made accessible. Finally, part of the archival material is still in his possession, either in his office, at his house, or on his computer. He has, according to the archivists, the prerogative of deciding what should be given and when, and of course he can always request that an item be returned after it has been given.

The Librarian's Assistant

The library assistant at the Moore is not trained to be an archivist, but she would like to become one. She is currently working with the head of the Moore on the Hawking Archive. This is on top of her normal duties in the library. She goes at her own pace, as there is no real deadline. She is just carrying out what the previous assistant was doing in the same way, as Hawking's students were carrying out what the previous students were doing. Like Hawking's personal assistant, what she likes about her job is "that she can do it on her own." When she has time, the head of the Moore brings down several boxes at a time and puts them on her desk. Then her job starts: "*It's completely random*, I know they [Hawking's wife and his assistants] are bringing more boxes over, and we are sorting through them, because not all of it is, I think, *necessarily of crucial importance for an archive*. So it's, um, ... we're just sorting out *until the interesting things can be preserved and presented* really. . . . I don't know what's going to be in the boxes."[44]

Contrary to Woltosz, who had to inhabit another self to create his programs (see chapter 1, n. 117, and chapter 3, "Throwing Things within the Universe from Infinity"), she draws from her own past and uses her experience to build up her categories and a coherent system of classification. Indeed, having been a student herself using archives (she has a MA in history), she thinks she knows roughly what is important. In addition, she follows her supervisor's directions in order to identify and write down the authors, publishers, dates, places, and so forth. But mainly, she just tries to "be logical for each box and work out what it all really is."[45] For example, she separates boxes that contain only one kind of thing (like box 20, which contains reviews of *A Brief History of Time* that have been professionally collected and photocopied by a press-clipping agency) from those that contain a more eclectic collection of items. "The ones that contain more things," she says, "I just separate them out into what *I* consider logical, umm ... categories, like reviews, articles, photographs, and so on. . . . *I really just have done it myself*."[46] But "what it all really is" is not always immediately clear, even when identifying one of the most well-known books on the planet: "Like this big stack of paper—for a long time I couldn't work out what it was, and it turned out to be the *Brief History of Time* manuscript [laughs]. . . . The chapters didn't have titles; it was a very early draft, and things were missing, so it wasn't immediately obvious what it was. . . . It was typescript, yeah, with someone's alterations on the side."[47]

In the meantime, more versions of *A Brief History of Time* have appeared: there's a rewritten first chapter with pencil annotations on it (but, as we know, Hawking couldn't have done it) and photocopies of photocopies, with marginalia, and spare pages, "which," as the head of the Moore says, "if you were being really brutal, you would just say: 'Oh, these are just photocopied pages.'"[48] The head of manuscripts will be able to look at the index and decide whether they are significant or not or decide if he wants to keep, for example, the many different versions of *A Brief History of Time* before the title was finalized. The other original copy is at the University Library, the only piece of the collection already there. Anthony Edwards—from Gonville and Caius College, who was appointed chairman of the University Library Syndicate by the vice chancellor—asked Hawking in 1993 to consider giving the original typescript of *A Brief History of Time* to the University Library so that it could be preserved in Cambridge for future generations. As he says,

> As you know, the majority of Newton's papers are in the University Library, which is the natural home for the papers of Cambridge's most distinguished scholars. But there is a further reason why I should make this request to a Lucasian Professor, because, as well as endowing your chair, Henry Lucas left his fine collection of books to the Library: "My desire therefore is that my said Books, such as they are, bee offered to the said University, and if they thinke fit, to allow them a place in their publique Library there." He left us 4,000 volumes, including Galileo's *Dialogo* and Harvey's *De Motu Cordis*. Should you feel able to accept my suggestion and allow *A Brief History of Time* to join this company, it would be a wonderful gesture to the University and its famous Library, and would be most warmly received and recorded.[49]

However, as the head of manuscripts says, "What we have is simply the typescript as it went to the publisher, so it doesn't differ from the published work; it's perhaps more of a symbolic significance than any scholarly interest, I would think."[50]

Though the librarian's assistant is confronted by the problem of the singular—so different from the final product that she can't identify it—she is also, like the graduate assistant, the journalists, and the social scientist, confronted by the problem of repetition.[51] The epistemology of the archivist is based on the notion of historicity—that one valorizes the uniqueness of a document. However, the material collected as part of the Hawking Archive is frequently characterized by its repetitiveness, by its iterability, as in electronic or published documents that can be reproduced at will *ad infinitum*. As far as the reviews of *A Brief History of Time* are concerned, she will leave them in their box: "They are so repetitive [laugh], there is only so much you can really say, I guess, about a book, especially if it is not a scientist reviewing it."[52]

And what does one do with all the speeches given in different places? Though slightly altered for each audience, they are essentially the same. It is thus difficult for the assistant librarian to trace their provenance and put them into context.[53] What would it tell us, then, about the "authenticity" of these documents if she could trace them to their place of creation when the computer and the journalists responsible for them are themselves responsible for—and reliant upon—similar acts of de-contextualization—that is, the copying and pasting of quotations or speeches and their translation to other contexts, as if they were pronounced there for the first time? "I find a lot of repetition, and it's hard to place where actually it was, and I kind of can guess by the date.... They were, you know [similar]; I'm sure he was giving lectures tours, giving all sorts of speeches a week."[54] Conversely, if the use of the computer makes her work difficult because of the immense amount of repetition it creates, it makes her life easier insofar as it allows her to distinguish the oral from the written, the speeches from the papers, because, paradoxically, Hawking has to write everything down, even that which is normally said but not written: "A lot of speeches, I couldn't tell where it was given or anything, but . . . they are clearly a speech before an audience, because he is saying things like 'Good Afternoon.'"[55]

As for tracing the origin, a similar problem occurs with the endlessly reproduced autobiographical accounts: "And I'd noticed he'd written a brief biography of his childhood which he put in *Black Holes and Baby Universes*, and which he then reproduced lengthier or shorter edited versions of, so I had six or seven copies of the same thing, but some of them were different; some of them have more, or less, . . . so I just try to organize it."[56] What does she do when she wants to differentiate these "similar" versions? She has to trace their origin and localize them; that is, she has to give them a proper identity. Sometimes she just looks for the date, or, if she has the date, she looks for the place. To do this, she uses Hawking's own Web page (maintained by Hawking's graduate assistant) or the web of knowledge; she also searches the University Library catalogue, which includes records of all Hawking's articles and books; the general academic Web site called Math Sign Net; or she looks at his books at the Moore.

Her careful gestures remind us of those described by Foucault when he talks about the rules constitutive of the author: "These aspects of an individual, which we designate as an author (or which comprise an individual as an author), are projections, in terms always more or less psychological, of our way of handling texts: in the comparisons we make, the traits we extract as pertinent, the continuities we assign, or the exclusions we practice. In addition, all these operations vary according to the period and the form of discourse concerned. A 'philosopher' and a 'poet' are not constructed in the same manner; and the author of an eighteenth-century novel was formed differently from the modern novelist."[57] Indeed, what we observe here is the construction of an "author" in the making. As the head of the Moore said,

> She was doing a lot of detective work, . . . getting an article and saying, "Well, where was it actually published?" Or there is something which, it looks like a chapter of a book, but she is going through all of the books in our collection to try to match it up with, is it a version of a speech or is it actually a draft of the chapter which is later in the book? Ah, so it's tricky, some of these things; it doesn't appear on these official lists of publications. The speeches can change from venue to venue, but not very much.

> So she's tracked down most things, but then you just have to say, well, it's a paper, and she can't tell me if it was a paper which was then published in [19]88 definitively.[58]

Later he added,

> She did a lot of thinking about things, when she was going through those particular boxes, about ... what this was, was it a version of the same thing? And so I think she got quite a good feeling for ... what it was; ... this was a version of something, or this was something completely different, or this looked—she would make a quick check with the prints, printed things, and say that "this looks like a revised version of this paper, and so it should be packed with this." And so that's why she moved to publishing [laughs].[59]

Thus, the assistant mainly deals with versions of printed materials that she identifies by opening books that contain roughly similar versions. However, the reason she likes to do this job is because, as she puts it, "I like getting the hands in the rough, the actual stuff rather than just reading a book. I quite like ... seeing the real material."[60]

In other words, when trying to match the versions she has with the ones she finds on the Internet or in the library, she makes visible the fact that the same material is already accessible somewhere else, but also, paradoxically, through this very same process, she also participates in the collective work of repetition by cataloguing this material and thus generating even more copies of the same. As I showed in chapter 3, Hawking uses neither pencil nor paper, so the back-and-forth work of scribbling, annotating, and drafting that typically makes up the work of writing and thinking is absent. For years now, he has had to do everything through the computer. This is true not only for Hawking, but increasingly for all scholars. Hawking's particular condition, once again, makes visible the practices of intellectuals who cut and paste from different documents to compose new talks, who recycle the same talk for different audiences, who transform their talks into articles or publish talks, interviews, and conferences as books.[61] In other words, the drafting process that would normally take us closer to the "origin" of the process of creation—that is, to "the author"—lead instead to a hall of mirrors. This is the case generally, but even more so with Hawking, who is obliged to repeat, copy, and cut and paste entire paragraphs from talks he has already given and papers he has previously published. The mark of the device becomes apparent in the construction of the text itself: "If there's a basic speech and some modifications, then that's very obvious. But I'm looking at this case—whether the speech, the millennium speech to the White House . . . included material . . . in his *Universe in a Nutshell*, . . . and then . . . it sounded very similar to that one that he gave at Cambridge during Science Week, the previous year. Ummm, but, obviously, it probably was a bit different in length, and so there would be different material in it."[62]

Though everybody has the option of saving drafts, it is very difficult to keep track of them or of the changes made in different versions of the same document. Additionally, many simply choose to "save," at which point the past of the document—its history—is elided. In this sense, as Lilly Koltun reminds us, "Digital data banks are not archives that have to do with the past, preserved like memories, succeeding upon done deeds and finished thoughts; these are archives whose future preservation is defined before they come into existence. They are not archives subject to selection, but to creation. Because digital data are so ephemeral and technologically dependent, they must be saved, if at all, at the moment of creation, or be lost."[63] As the head of manuscripts told me, putting into sharp relief the impact of digital media on the meaning of historical record, "Most people when they correct something in, say, a word document, . . . changes [are not] recorded, you can track changes of course, but other people don't normally do it like that, they just correct."[64]

Thus, Hawking's signature emerges—thanks to the hard work of the librarian—through the combination, superposition, juxtaposition, and reorganization of the "immutable mobiles."[65] Like the scientist or the ethnographer, the librarian produces knowledge through the mobilization and manipulation of inscriptions, reminding us that, indeed, the archive is a center of calculation.[66] By juxtaposing Hawking's different "speeches," the librarian makes visible their similarities—entire paragraphs are reused—but also their differences: the same paragraphs are used and reused but for different reasons and in different contexts; rewriting is made evident to her eyes by the varying length of the different versions of a speech.

So, what happens if Hawking gives a similar talk with a different date? Is the document still considered an "original" in the sense that it was produced in—and for—a different context?

> It has been printed out. Probably it is identified just, you know, as "the speech given at the Science Week." . . . Well, it's a document. It's proof that he gave that speech in that place, or allegedly [laugh], *that was the plan, anyway, to give that speech at that time in that place*, and so, that's the evidence. . . . And so it's difficult to know what the actual performance was, and if we don't have access to it—a digital tape of this, this is exactly what went on, and . . .[67]

"Originals" will be stored in the University Library, where eventually they will be copied, digitized, relocalized, and displayed in the Hawking Archive at the Moore or in a digitized exhibition, which in turn will one day be accessible from every computer.[68] Indeed, as Mike Featherstone argues,

> The archivist, librarian and professional researcher create the maps and record the journeys into the archive that produce the images we have of the possibilities of the material. Yet such classificatory schema and mapping devices can disintegrate under the volume of inchoate material, which threatens to defy the impulse of order. This is the image which Borges elaborates in his short story "The Library of Babel," a library in which all the books in the world in multiple translations are housed in an infinite number of galleries. A library in which it would be possible to destroy millions of volumes, yet the almost identical material would still be available in millions of others.[69]

But what is the "real" material here? Perhaps the correspondence? However, the assistant will have to identify its authors and by this process extend her own competencies: "I look at the people he's writing to. Just put it onto Google, just to have a better idea of who they are. But that's linked to my interest, it's actually more than necessary, as long as you get a name, that's okay with some of the most legible handwriting; [or] I look at the Internet to try to find out how to spell their names; because I can't understand a letter. Yes, to have a rough idea of who he's speaking to."[70] And if she can't find anything on the Web, she asks the head of the Moore or one of the librarians: "There hasn't been anything I have been completely lost with, that's, as long as there is a record of everything, however brief, that's not really my job to find—to trace the origin."[71] Or, if she has no date, no place, the only information she will give is that she doesn't have any information to give, for example, regarding the speeches. The question is do they become unique—originals—because she can't match them to anything else?

> I just wrote down the most information that I had . . . [for some] I put they have an unknown origin, or an unknown date, because there is no way I could find it . . . so [I] just try and give as much information as I possibly could, but when they don't have a date, and they don't have a place, I can't do it, I mean, that's something that a proper archivist I'm sure would be able to do because they would be able to go and ask and go to Mr. Hawking himself.[72]

Paradoxically, though she says it is easier to work on Hawking rather than on another scientist "because she knows who he is and his work"; it doesn't make any fundamental difference that he is next door. She doesn't ask Hawking when she doesn't know. This is the head of the Moore's responsibility, but also it probably would take too long to get an answer. She has met his entourage and knows that she "could" talk to his graduate assistant if she had to, but it has never been a necessity: "If there is something I didn't know, I could ask him [the graduate assistant]. Really, the resources are here. I mean it's just there. It's not difficult."[73] She has to trace what's there and make a brief list of everything before it is accessioned; again, she is not really supposed to find the origin.[74] Thus when she says, "I know him and his work," it doesn't mean that she knows him personally. It means that she has read his book *Black Holes and Baby Universes*. And it is according to this standard that she will select and organize his writings.[75]

Interestingly, Gribbin, who is the co-author of *The Life of Stephen Hawking: A Life in Science*, never met Hawking while writing his book, but was given the drafts of *Black Holes and Baby Universes* by Hawking's secretary. The reason he never talked to Hawking in person is that Hawking had already given an interview to another writer, whose book Gribbin reused as well. Thus, on one side, the assistant uses the book as a standard to localize, make an inventory, and organize Hawking's "previous" writings; at the same time, she once again produces more copies.[76] On the other, Gribbin uses the book as a base to create another text that will participate in the standardization of Hawking's biography.

> HM: But do you think, because he's a celebrity and because he's next door, for you, do you feel that it is more difficult to do this work, or it doesn't change anything? If it were Newton's papers, would it be the same thing?
>
> J: I think it's a little bit easier for me because I know very roughly what it is that he's done.... If he was a scientist that I had no clue about, I'd find it very difficult to recognize work and what stage of his life he's at. I find it quite interesting; I read *Black Holes and Baby Universes* just to ... so I know, roughly what it is that he'd done, ... but no, I don't think that it makes it harder; ... it makes it easier because I know who he is. No, I don't think the celebrity ... I mean, I knew that he works here when I came here, and I think I saw him on my very first day.... I was quite pleased to see him, and after that it just settles into your normal life really, so I really don't think about it in those terms anymore.[77]

But she won't have the final say as to what should be preserved or not: "Because I'm not ... I don't have a scientific background. I wouldn't really know what should be included, but ... I think I have enough responsibility in just seeing what I have and trying to organize it the best, but I wouldn't be comfortable saying, 'You can throw this away,' because I wouldn't know that yet, but it hasn't got to that stage."[78]

As we saw in the first chapter, the graduate assistant was responsible for narrowing down the journalists' questions by eliminating those that were similar, allowing Hawking to respond to a number of analogous questions at the same time, questions, one should note, that have been asked—and answered—over and over again whenever he consents to an interview or when he gives a talk. So what is an archivist to do? What is she supposed to do with the thirty similar copies of an article, or with a response to a question that has been asked and answered many times, with only slight changes of wording?

> I don't know that, I don't think we like to throw away anything. I've noticed that just from working in the library generally, but there must be some things in any archive which are not that outstanding. . . . I don't know how you'd make that judgment; *maybe they just wouldn't be on display*. It's hard because there might one day be someone somewhere who is interested, so there's only so much, you could, spacewise, keep everything; . . . that's pretty much what you do, . . . but I've noticed that there are some things . . . thirty copies of the same article, do you keep it? Things like that. I have no clue. . . . I think . . . it would be terrible to throw something away, and [then have] someone wants it.[79]

Someone like a sociologist or an anthropologist who might be interested in analyzing the process of repetition at play in the construction of an identity. *Identity* has an interesting double meaning here: the identity of the collected texts with each other and the cumulative sense of the writer's identity as a constant behind the variability of expression. But, paradoxically, they are dealing with a living person here who is still producing material. In what sense, then, is something new produced, and what does "collect everything" mean in this particular context (if collecting everything is possible): "I mean, with a living person [who] keeps producing lots and lots of copies around the world, it's going to be hard to actually collect everything. So I don't think we can claim to be actively collecting everything, but essentially just to keep the material which *he* thinks is relevant at the end."[80]

This assessment points to the presumptive importance of Hawking (and of the funding to collect and maintain his archive), so that the volume of material seems to be a minor consideration. Indeed, though the head of the Moore has no intention of insuring that they will get a copy of all Hawking's old and new material, he still collects copies of reports of major events, published for example in the *Cambridge Union News*, or in other newspapers that come to his attention (e.g., concerning Hawking's zero-gravity trip). But, mainly, "it's largely a lot of accretions; . . . we don't know if next week, somebody might say, 'Right, we're clearing out again. And we need to deposit ten boxes with you. Are you happy to take them?' And that could happen at any point."[81]

As for the personal correspondence, how is the archivist to know what will become important? What basis is there for a decision to keep or discard an item?

> It's hard . . . because then, of course, I see . . . the letters that Hawking has written when he's a young sort of up-and-coming PhD student or lecturer. He's written to older professors for their advice, and you actually don't know who the young people are who are writing to Hawking, who they will become, because they might become so important in their own time that a letter that they wrote at twenty-one years old is then more interesting to future generations, I mean, I'm sure he does get bombarded, he must do, with people asking for advice or so on, but these people could turn out to be quite interesting themselves, so, . . . I mean, I haven't seen too much of that; most of it is private correspondence, people he obviously knows and stuff like that, but he must get so much attention.[82]

However, when I interviewed the assistant librarian, there wasn't much in the way of material to be archived. What they had, more or less, were the same articles, books, and magazines that could be found anywhere else. Though there is still a sense that his identity resides somewhere else, they believe a more complete and coherent picture of who he is will eventually come into view: "*I would like to know more about him* and tackle more. . . . I feel like, that just the little pieces that we have, that we get through and through the picture of it. I guess that would build as it comes along."[83] Thus, like the students filling up Hawking's silences on the stage so as to create a sense of continuity and a unity, the librarian will fill in the gaps in the construction of the archive in terms of chronology. Historians will thus, one day, be able to tell their stories—that is, they will, in turn, fill in the gaps, frame and put into context, complete and create a unity.[84] As the librarian's assistant put this, "The story I think is best . . . left to the academics, to the people who look at the archive."[85] In reassembling Hawking's material in a certain way, one creates a path to tell a story. By selecting, cataloguing, and preserving "his" material, each actor participates in the construction, maintenance, and extension in time of Hawking's sacred body.

THE ARCH-ARCHIVISTS (THE UNIVERSITY LIBRARY)

The head of manuscripts at the University Library is used to taking care of papers left by people who died hundreds of years ago. This is not the case for the curator of scientific manuscripts, who is approached regularly by people asking how to leave material, how they should keep things together, how they should file things, or what they should keep, though nobody did this for Hawking. Moreover, it seems that the different parts of the collective body at the time of my fieldwork hadn't yet begun to talk to each other: "You understand that one arm doesn't always know what the other arm is doing. . . . And . . . nobody's consulted me about the Hawking collection at all. As curator of science manuscripts, . . . it's rather curious, . . . but, as I said, I know nothing of its content, at all. . . . It's deposit, accession, when did it go there? Who sent it? How did it go there? Who took it? [laughs]."[86]

A few months later when I came back to talk to the curator of scientific manuscripts again, I realized that my presence had triggered his curiosity. With the head of manuscripts, he had gone over to the Moore to talk to his counterpart and to look at the Hawking material. More boxes had arrived, and more "archival material" seemed to be present. As I was talking with him, he had just received the final version of the inventory made by the head of the Moore. It was now in his hands, and he was looking at it for the first time with me as I was asking him to reflect on the different items. Another process of selection was occurring, and some of the questions raised earlier by the assistant seem to find their solution.

For the curator of scientific manuscripts, personal papers—what people have kept about themselves—obviously matter. Press cuttings, for example, are of value: "If they . . . the family, or whoever had said, 'Oh, we must keep that article,' snipped it out, and kept it, we'd keep it. We wouldn't think of discarding it."[87] The press cuttings have value in "person-terms":

> Somebody's snipped around the paper to . . . take out the cutting, I mean, so, *it's actually, physical, you take action* to extract this information, which would probably, in the pre-computer age—I mean, nowadays—when you have press archives, you could go to the *Times*, two years ago, couldn't you? And you could search an edition of the *Times* for a certain name—you could look for Hawking's name. . . . But, before [there were] those archives [when these materials were presented in this way by somebody], . . . it would be difficult to find that information just by thumbing through a newspaper.[88]

Contrary to what we have seen previously, where the definition of an action was seen through its very absence (the fact that material hadn't been thrown away meant that it was wanted), here the trace of an action is left by its physical presence (the mark of a pair of scissors) that allows us to read the "intentionality" or "the will" of its author by creating a link between the material and the person "that extracted it."[89] It is then perceived as his possession. However, if the person herself can physically do the action, it can also be done through intermediaries: the entourage or a press-cutting agency. Though the work of selection, classification, organization, and mobilization of the material is performed physically by others, this collective work doesn't change the nature and the definition of the "personal action." What matters, as the curator of scientific manuscripts says, is that "*you* collected it, and it's about you."[90]

Thus, on one side, press coverage of all the articles that have been collected around one topic (for example, box 20, containing reviews of *A Brief History of Time*) or translations of the same article in different sources are kept together in one place as part of the "Hawking collection." Not only are they perceived as being part of his personal collection for the reasons mentioned above, but they are also kept there for the sake of convenience. Indeed, even if the material is available at the University Library—since all the journals, magazines, and newspapers are already there—it would be easier to consult them in the Moore: one wouldn't have to go through the whole newspaper to find an item because it would just be there as a cutting. On the other side, the same book might be duplicated and redistributed in different libraries. For example, the University Library has one copy of Hawking's Sixtieth Birthday Conference on open-shelf access and one copy in the special collections. They may decide that they want a second copy to go into their "Cambridge" class file as well, because it's about Cambridge and the university. The Moore has requested three copies from the Cambridge University Press—one for the Central Science Library, because it's of general interest to the science community; one for the Moore as a circulating copy, because it's obviously relevant to a math library; and a third copy to be deposited in the archive. As the head of the Moore says, "We've got a spare copy that's not going to go anywhere, and we're not relying solely on the one which is in special collections."[91]

The press clippings will be kept together insofar as they reflect an individual's desire to collect material about him- or herself; similarly, what has been tied together (compiled, for example, in a folder) will stay together—it will be considered an inseparable, integral whole by virtue of the individual's desire to classify them—collect them—together. Each item thus finds a context or a series within which future scholars—so the archivist presumes—can interpret it.[92] As part of the unpublished material, it will be kept at the University Library: "I mean, obviously I'm interested by these letters from the publisher Bantam . . . and a letter from Harmony Books. If there's . . . one file folder, um, whatever, kept together, in some way, I wouldn't split the folder up, . . . because it has some integrity, because he's kept the thing . . . together."[93] And, as he mentioned earlier, "It's a general archival principle, I think . . . you will find it generally adhered to, that . . . you keep connections together. . . . You don't try to separate things."[94] A rough draft, letters addressed to important people, letters autographed by Hawking, all these pieces will be preserved at the University Library: "You see, the rough text of an article is probably of some interest; it's kept with this, um, about his medical condition. . . . But these are, they're significant people, and . . . names like that just . . . jump off the page, like they're in bold type. . . . No question. [Laughs]. And, . . . so you see, it's him himself, writing to Wheeler [at] Princeton, I mean this is just two, typed copy and written copy. . . . Who handwrote it? The secretary, probably."[95]

Will Stephen Hawking's reliance on his extended body have an impact on the ways things are preserved? And how can Hawking claim his authorship? Interestingly, there is a sense in which one follows a form of disincorporation of Hawking, not because he is disabled, but because he is a scientist, for the state of the scientist's body has no impact on the ways theories are produced—they are the products of minds, not bodies: "It's scientific endeavor, you know, it's observation, recording, theorizing, hypothesizing, if you like, creating serious mathematical work, and that's essentially what Hawking's doing, and we might say Boyle or Feynman, who were both hale and hearty, nonetheless, were working toward the same goal in physics and cosmology. And really, it doesn't affect physics or cosmology, the fact that one used to walk for dozens of miles every week, and Hawking can't."[96]

Moreover, according to the head of the Manuscript Department, if material is generated as a result of an assistant's relationship with Hawking, it should be part of the archive. In the same way, for the curator of scientific manuscripts, the person of Hawking doesn't end where his flesh-and-blood body does:

> People like Ernest Rutherford, I mean, he was active and in good health for almost all his life, and was writing and speaking, and we preserve his writings and his speaking, his notes and his correspondence, and everything, if you look at those collections. . . . It was just the sheer odd chance that Hawking has been afflicted in this way. . . . It shouldn't really affect how you keep, try to reflect his life's work. And . . . if you try to impose a restriction,

> to say, well, okay, because it was Hawking who dictated a letter, but actually it was written by, you know, somebody else, and say, "*That's where it's not Hawking*." But I don't think that, really, I mean, you have to interpret it broadly, because, this is, as far as I know, . . . we're probably not a unique case, there must be other scientists, and individuals, and people who we'd want to keep collections of, . . . and who must, similarly, have been in very difficult circumstances like that. . . . But it is unusual, isn't it? . . . *I know of no other case, but it just happens he's a scientist, and he's in Cambridge*, and so, it's come to us. And, there must be, poets, and I don't know . . . [97]

The curator's comments raise the question of attributing works to Hawking that were written (whether by pen or keyboard) by someone else. But again, this is not unusual in either the history of science (e.g., Boyle's and Newton's use of amanuenses)[98] or the contemporary practice of science (see chapter 2 above).

For the head of manuscripts, the Hawking collection is the first he has dealt with in which the substantial proportion of the archive is in electronic form. For the curator of scientific manuscripts, the fact that Hawking relies so much on his computer doesn't change his approach.[99] It makes visible a common problem that will occur for everyone in the computer age, regardless of whether they are disabled or not. Archivists know they can keep paper for decades and centuries, because they have done it, and the fact that it is there at the University Library proves it. What about electronic material? How are they going to make this material accessible? How are they going to keep it? How are they going to make sure that it's going to be available? There is no policy to print material out, for instance, to take a CD-ROM and print everything that's on it. All these questions are still unanswered. The fact that Hawking is a scientist makes his case even less unique, since scientists were the first ones to use software and digital apparatus:

> Everyone is dependent on computers, certainly—however able-bodied they are or otherwise. Um, and so there's going to be a huge amount of unique material, kept in only one place, essentially . . . people's hard drives, with their e-mail correspondence, or something. . . . What on earth are we going to do with all this? But, that's the same for Hawking, and it's the same for the guy in the next office . . . who plays tennis every night and . . . does handstands for a living, or whatever. So, I think you have to try to establish broad rules that can be applied . . . to all these cases.[100]

What the curator of scientific manuscripts will keep as scientific manuscripts, in the broadest sense, can include all sorts of magnetic media that

Hawking might have personally generated: correspondence, papers, written material, personal creations, unpublished works. They won't keep offprints or reprints. One can't be too narrow about the definition of "personal creation," for again something could be produced through intermediaries, secretaries, assistants, or nurses. In general, however, personal creation is thought to be a product of human action rather than of a machine that can print hundreds of copies:

> You can't be too rigid about this. But in *this* case, because he must have done so much work through his secretary . . . and his nurses, . . . anything like that [which indicates] a human, rather than a machine printing . . . out, hundreds and hundreds of copies, [like] a library collection. So when somebody has put something individually unique, if you like, that's the sort of material that I will be looking at.[101]

This view is difficult to reconcile with the kind of material that Hawking is producing. First of all, his speeches are reprinted hundreds of times by a machine. Shouldn't they be part of the personal creation because of that? Second, archivists are looking for something unique. In other words, the value of an item resides in its individuality—in the fact that it is the only one in existence. It is exactly the opposite with the items that Hawking produces, where the value of an item is based on its repetition. What makes a paper valuable, in this sense, is not the fact that it is "unique," but the fact that it is reproduced and cited a million times; and conversely, it is because it is reproduced and cited a million times that the aura of "the first draft" grows.[102]

In the same way, a photocopy of an article is not interesting for the curator of scientific manuscripts, who is looking for "unique items," unless the original has disappeared or if it might be useful to future scholars insofar as it indicates that "[Hawking] wanted it." So, it will be kept at the Moore: "There are storage facilities for press cuttings and individual papers like that, so . . . there's quite a lot of that. . . . You can't know that the original hasn't been destroyed, in some way. . . . But if there are thirty copies of a photocopy, I probably would destroy twenty-nine of them." And as he continues going through the list, he says,

> So it's, you see, *this* (he points to an item), there we are, you see, *this here* is his ground paperwork—very likely that's of some significance, and it's not published, so it would be something we'd keep. . . . General correspondence, general correspondence, yes, yes, yes, yes, so this is box after box there.[103] . . . Handwritten notes and typed scripts, there's no question we'd

> keep that. . . . [Pictures] of Hawking, himself. Looks rather nice, doesn't he? Children, colleagues, . . . *Star Trek* cards! [laughs] . . . Oh yes, yes. It is very personal, isn't it, anyway? . . . And there's not going to be—*no other copies of that material*—anywhere, I shouldn't think so. And that's where we come into our own, where we can actually help with things, where we can make sure that sort of material is catalogued.[104]

And again, photographs, along with paper records, manuscripts, typescript, examples of handwriting, that embody "work in the making," are also of interest to the archivists:

> Are there any photographs of the inside of the office, or anything like that? On how he actually worked and operated—I mean, there probably are, and if there are, the photographs would be of interest to us; we would keep photographs. Let's say that they're not unique. . . . If you've got a negative, you could always print more, but, usually, that, a black-and-white photograph is practically all that you'd find, in personal terms. The printed ones, . . . probably of your holiday, . . . you've taken the photo, you've got one print, and that's it, isn't it, then? [. . .] And you probably lose the negatives. And then that's the permanent record, isn't it? And so that's what we do.[105]

EXHIBITING THE DIGITAL: THE HAWKING ARCHIVE

Material considered "authentic," "original," "unique," and "unpublished" ("drafts," grant proposals, photographs, personal correspondence, handwritten letters) will be housed at the University Library. The collected published material, on the other, will stay at the Moore (books about Hawking and by him, magazines articles, and maybe cuttings, which are being debated). As far as Hawking's electronic material is concerned, this is a question that still needs to be addressed. It might be taken and administrated through D-Space, that is, a specialist repository for electronic material. The goal of the library's digital archive project is to digitize administrative or personal archives as well as articles associated with Cambridge University to create the Cambridge Institutional Archives. As the head of the Moore says, "Having a subset of that which is Hawking's materials has been something which is quite down our street as well."[106]

The archive will be composed of two halves: the digital archive, which is basically for future generations because it's like a virtually closed archive guaranteeing access to material in a form that can be converted from technology to technology; and at the front end, a kind of broadcasting of archival material, as the British Library does. "This is very simple to do," the head of the Moore Library adds.

> It's a kind of adding metadata and all the stuff to insure you described it. To an extent, it strips out bits rather than straight digitization, and on our existing platforms, which we could do, so there's two sides of it: keeping it forever and then broadcasting through the Web.... The project is in progress; they are currently looking at getting money for the Newton papers. Well, the papers are fairly on the way, but there would be a similar process of digitizing all of those. The University Library again has lots of experience in the current digitizing project. And there is a lot of material around the university and the colleges as well that they [will] do. So as the cost of the storage goes down, [and] then we are gaining in experience, so if this material, which is suitable for that, then we do that, and then it makes any of our exhibitions become . . . have a digital end. That you don't have to go to the library to see the exhibition anymore.[107]

The idea is to use Hawking's material to promote a general project on digital archiving and to find sponsorship for it. At this point, there is a plan to display this material as a pilot study. The head of the manuscript department knows now what is there, and, with permission, the Moore hopes to put on a digital exhibition. For the head of the Moore, the Hawking Archive Exhibition could be thought of as a *"figurehead of an exhibition,"* a means of attracting donors. As he says, "If they want to fund a digitization project, then digitizing Hawking's material could be quite significant.... Because I mean, he gets a lot more press coverage than Isaac Newton! [laughs]."[108] Indeed, he thinks that a substantial project could be funded if they could just sort out questions concerning copyright. They could then scan printed material, such as the articles in foreign newspapers from China, India, or South America that are the most delicate, because they are not on terribly good paper: "So you have books, the honorary degrees, . . . medals and things on display, and you could browse through articles, things, and so on, and digital copies of general newspapers with particular links by page.... You could do that from everywhere; *it's a good selling point for the university.*"[109]

This material could be available as a resource, as a way of tracing Hawking's presence:

> So if you were saying, right, you know, he was in Mexico, and then he was in Venezuela, and back to California over a certain period, and then you could see that he was in India, at a certain point, and, . . . my index

> says, right, there's a picture of him visiting an ancient observatory. So we know part of his action; we don't have the actual itinerary in the archives. I mean, this is what we're doing, this is the way we are using the dates that sort of ties everything together, if there are gaps. So you could pop around the world with these newspapers.[110]

According to Michael Lynch and David Bogen,

> The recent proliferation of electronic means for reproducing and disseminating documents and entire archives has begun to disrupt the traditional exclusiveness of scholarly access. Derrida's "institutional passage from private to public" is no longer contained or "domiciled" in a site, the visiting of which may require long-distance travel, scholarly credentials, and connections. A Web site may be difficult to visit for persons who do not possess or have access to the requisite technology and skills, but together with other forms of electronic media, it has the potential to turn a body of documentary evidence into a "popular archive" subjected to mass visitation, reproduction, and dissemination.[111]

In the same vein and paradoxically, the exhibition, as it becomes permanent, would also become much more interactive and changeable.[112] More than a simple Web site—like Hawking.org, where published papers and detailed information about Hawking are already accessible—the exhibition could turn into an attraction where people could trace Hawking through his talks and through different newspapers, or have access to different versions of his manuscripts (e.g., *A Brief History of Time*). One could even track the different versions of his Web pages, or the user (the public) could add more material—to fill the gaps and complete the picture—if it was available:

> Since we're in a kind of a very international section of Cambridge, . . . and the Maths Institute [is] next door, so there's lots of people from various countries, who are constantly coming through, and . . . the turnover is much faster than at a "general" university. . . . That's one reason, you know, from time to time, we will get another edition of somebody bringing in, and saying, "Well, the last time I saw this archive, here's a new edition of the copy I picked up." So, it's actually striking a chord with visitors, that, if they came, and said, "Right, let's see what happened the last time he was in my country," and, bingo, there's an article [laughs].[113]

More than a trial balloon, the Hawking Archive would become a black hole around which scholars would naturally gravitate:

> Well, the more you go along the lines of saying, yes, we want to digitize it, and make it available to the public, then . . . it just increases *the internal gravity of it*, and so it would attract more things coming in, which is neither a good thing or a bad thing. To say, okay, now that's it's turning into a place where all the material on Hawking is in one place, and therefore Hawking *scholars in the future would naturally gravitate* to come here, to see the library, because they've got all the material, and they've got material by him (pause), about him in/at the same place.[114]

One could also imagine that a museum could work in tandem with the exhibition: "An internal memo mentioned that there is a room where his previous wheelchairs are stored. . . . If we accumulate a lot of material, which is not really long-term compatible with the library, then, if we have a partner museum, the two things can work in harmony really."[115] Or we could imagine creating a virtual link between the museum and the digital exhibition.[116]

Thus as soon as the Hawking Archive/Digital Exhibition/Virtual Museum is completed, it will assemble and centralize in one site items such as the different versions of Hawking's manuscripts, conferences, speeches, letters, Web sites, . . . wheelchairs (?), or even maybe his synthesizers (that could even be played!); and all these collected digitized objects will be accessible from any computer on the planet, as a unity or as bits and pieces that can be rearranged, sorted, and recontextualized, as part of the continuing work of exegesis.[117] As Lilly Koltun has argued, "Ultimately, what is revealed by digital options in their new impact on archival actions is an older, but unminded perception: archives expose not a finished past society, but, as always, and as Raymond Williams pointed out, our own, present society which does the selecting, the keeping, the using, and the constituting of its many stories in continuous and contingent re-making."[118] Gordon Moore has not only made it possible for Hawking to use a computer, an archival device, but also, through the Moore, he has given him a site to centralize his work and (personal) possessions that will be digitized, redistributed, and made accessible to countless interpretations.[119] In what sense is this archival body his body?

⁕

> HM: Hawking uses this machine to talk. He has a program with his own vocabulary and [different devices], so I was wondering to what extent [this machine] belongs to an archive.
>
> P (the archivist): Yes, I mean, that's more a museum object [laugh], perhaps, than an archive. [120]

Einstein's brain was preserved, dissected, and subjected to various efforts to define how it differed from a "normal" brain.[121] Could we imagine a future archaeologist of Stephen Hawking's computer exploring its different levels, its many layers of sediment, its tools, its refuse?[122] Henry Lucas, the founder of the Lucasian Chair of Mathematics, left four thousand books to the University Library. Might our imaginary archaeologist be able to discover the "four thousand vocabulary words" that Hawking frequently uses? Did Hawking, or anyone, keep track of the changes he made—the proper names that he added or removed that testify to the different places he went and the different people he met or worked with over the years (e.g., the graduate and personal assistants, students, and colleagues)? Will any trace remain of the always-changing socio-technical networks that surround—and enable—him? And what about all the stories he tells about himself, all the comments and quotations that have been stored in his computer or the computers of his assistants, in response to the oft-repeated questions by a constant stream of journalists? And what about his favorite insults? His hobbies? The family matters that should be accessible, thanks to the filing system in EZ Keys? Did he create different categories from those that were offered to him, like personal needs, favorite foods, or jokes, and so on? Perhaps the archaeologist will be able to guess from the complicated and outdated systems on his carefully preserved and archived computers how Hawking could recall, across all the files stored there, the one specific file he desired. Perhaps she will conclude that, for Hawking, finding was more difficult than creating anew. This perhaps explains the instability of the original not only on the computer, but also in the archive, with all its drafts, copies, repetitions, and redundancies.

Will the archaeologist have access to the Equalizer folder—the "one file" filled with plain text files—that, according to his assistant, contains everything that is important to Hawking from his physics to his family, everything he talks about and everything he does, including his lectures and the various speeches he has given. Backed up religiously to other safe locations in just a few seconds, it is very small, though if it were printed out it would fill up an entire room. Will she be able to recover his articles, his lectures, his programs, even his voice on his assistants' computers? And as for the diagrams, will she find any trace of their ephemeral presence outside his students' papers or his brain, once they have been erased from his blackboard?

Indeed, by looking at the structure of the screen and the organization of vocabulary, will she be able to understand how Hawking's brain worked? Contrary to the archivists who followed traces left by a pair of scissors or a string surrounding a collection of papers or a box, will she be able to guess

Figure 7. The Hawking Archive. Photo by Lauren Kassell.

what files were most important to him—because what he used more often tended to start with the letter *A*, and what he used less often tended to start with later letters? Indeed, it is easier to open files that were near the beginning of the alphabet. Perhaps, like the archivist, she will be able to conclude that what wasn't deleted was also what he "wanted" to keep. Later on, perhaps she will be able to understand how the brain in general works (the brain and the eyes are able to process more easily groups of three sentences than four; horizontal displays of words are better searched by frequency than alphabetically, etc.). Perhaps, through the computer, she will be able to understand how we thought about cognition—as distributed. And perhaps, again using this electronic relic, she will be able to analyze how language functioned in Hawking's time: for example, the thirty-six most frequently used words in English are at the bottom of the screen. Or, if she were to find my interview (see chapter 5) and then trace it to its place of publication (under your eyes), she might be able to understand in what sense objects mattered to sociologists.

But maybe Hawking will decide that all this should not be retained or displayed at all. As far as his assistant is concerned, "It wouldn't be a good idea to give his computer *now*, because there's a lot of information on that, that *we* probably don't want the whole world to see at the moment, but it would be a nice thing to do at some point."[123] Without all this information, what has been "archived" so far is the persona of Hawking: published books by him, magazines and books about him, and the typescript of *A Brief History of Time* as it went to the publisher. The rest of the boxes are still waiting in the basement of the Moore Library. In some of the boxes, I even found a few pictures of myself. And indeed among the many books currently on display at the Moore, one can see *The History of the Lucasian Professor of Mathematics: From Newton to Hawking.*[124] As Harriet Bradley reminds us, "In the end, what we hear is not, perhaps, the lost alterity; above all, what we find in the archive is ourselves."[125] Here, the production of the ethnographer has literally become a part of the archival body (see fig. 7).

VII The

Hawking Mee

I begin by drawing your attention to
curious feature of the notion of doing som
do something. In the end I hope to satisfy you tha
than merely curious; it is of radical importance for our merely
tion, namely, what is *le Penseur* doing? ing to

Gilbert Ryle, "The Thinking of Thoughts: What Is 'Le Penseur' Doing?"

For the philosopher, it was a pure experience in thought; for the ethnologist, it became a scene as beautiful as it was improbable: the encounter between the man Hawking and his plaster statue—The Thinker. The scene took place in the professor's office in Cambridge at the Department of Applied Mathematics and Theoretical Physics (DAMTP). There a statue representing Hawking was presented to him so that he could give his verdict before its final version was cast. From plaster, it will be made into bronze and then inhabit the gardens of the Centre for Mathematical Science at the foot of the professor's building. For the occasion, Stephen Hawking, his assistants, two physicist colleagues, the sculptor, the statue, and the ethnologist (I, the author) were invited.

In this chapter I describe the interaction that took place between these actors. I will take into account (1) the materiality of the statue as it moves from plaster to bronze; (2) its circulation (it moves or causes the group to move during the interaction; it will also be imagined and projected in thought to other places within and beyond the building while simultaneously being endowed with different qualities); and (3) its presence and what it allows (a sort of collective "*délire*," where the actors—including Hawking himself—make joke after joke comparing the "true" and the "false"). I thus endeavor to follow the genius being made present, put into words, and shaped. Where is HAWKING? Where is the real one? And where is the copy? Where is the original? Where is the replica? Who is who? Who is doing what? What is/are Hawking, The Thinker/The Genius, this immobile voiceless man and/or statue doing? These are the questions that this chapter attempts to answer by providing a 'thick description' of this scene, to use Clifford Geertz's term, which, we should not forget, was inspired by Gilbert Ryle's "The Thinking of Thoughts: What Is 'Le Penseur' Doing?"[1]

[illegible] way subjects [. . .] construct the object. [. . .]
[illegible] other aspect of the story: how objects construct the
[illegible] second half of history are constituted not by texts or
[illegible] remainders such as pumps, stones and statues.
[illegible] ave Never Been Modern

[illegible]ng started with a harmless e-mail sent by the Department of His-[illegible] and Philosophy of Science at Cambridge, from which I learned that IWC Media limited, a Scottish company, had decided to make a documentary about Stephen Hawking to commemorate the twentieth anniversary of the publication of *A Brief History of Time*. For the occasion, the filmmaker wanted a student to cross the courtyard of Caius, Hawking's college, with the famous book in her hands; later she would be filmed rowing on the river Cam with other students. For this purpose, they wanted a young girl of about twenty-one, with a fresh and natural look, someone who would not be too much of a contrast in terms of style and appearance. I immediately assumed that Hawking would also be filmed, although there was no mention of this in the e-mail. I called the movie director for permission to observe the shooting. An assistant answered . . . and very quickly slammed the phone down. I learned that the director was already rowing on the Cam, and that it would be impossible to accompany the crew while it filmed Hawking, as his office was not big enough. "Rather talk to God than to his saints," I said to myself, so I decided to go over the head of the director's assistant and contact Hawking's graduate assistant directly, who, after weighing the pros and cons, went for the cons and called me back to refuse. However, he said, as if to give me some consolation that "they" [the assistants] thought that I might be interested in participating in an event that was to take place the following Friday. The sculptor Eve Shepherd, who was already being compared to Rodin, would be coming to present Hawking with the statue she had made of him.[2]

FIELD NOTES: HAWKING MEETS HAWKING

The meeting is scheduled for 3:30. I ride my bike, and pass in front of the DAMTP; Hawking waits in the rain. His nurse closes his van. I say hello, but he doesn't see me. I climb the stairs, while he takes the elevator. By the time I arrive, he is already there. The PA [personal assistant] is at the door talking to the sculptor; they are in front of the statue of Hawking enclosed in bubble wrap. The PA wants to take a picture. The GA [graduate assistant] makes a joke: “The statue is still wrapped in plastic.” The PA, seeing that I’m waiting, wants to include me in the conversation and says to the sculptor that I’m here . . . “to observe.”

The door opens. The GA carries the sculpture into the professor’s office. He puts it on the floor and removes the bubble wrap that protects it. The statue of Hawking in his wheelchair sits there. The PA, the GA, the sculptor, her companion, Hawking, two physicists who were walking by, and myself look at it. Before the GA brought the statue into Hawking’s office, the sculptor told me that it was an immense amount of work to re-create the wheelchair. She takes me as an example: “If I had to do a sculpture of you,” she says, “I’d have to take all your measurements, but here I have to integrate the measurements of the wheelchair, which is very complicated. This is why I did the head separately.” The sculptor also mentions something about the resistance of the material: “You can’t do what you want with it.” Apparently this is the first statue to represent Hawking in his wheelchair.[3] It was ordered by the department. I was told that the statue will be displayed in the garden below Hawking’s office in front of the DAMTP. From what I understand, it is only “the first draft” that is being presented to Hawking today; the final version will be made in bronze.

Hawking doesn’t seem to be able to see the statue very well. He moves his wheelchair toward the sculpture. The statue sits in front of him. But the light coming from the window behind makes it difficult to see. He asks for something “black.” The GA takes down the professor’s black gown and puts it behind the statue. But this is not what Hawking wants. He wants the statue to be moved in front of the blackboard. The GA adds that it’s normal after all: “It’s always like that, Hawking always has his picture taken with the blackboard behind him.” The statue faces toward us. Hawking faces the statue. The sculptor explains to the group, which keeps asking her questions, how she made the sculpture and that she worked from pictures she took of Hawking, and she talked a lot to another sculptor who previously made a statue of him, but she mainly works “from life.” Hawking mentions something. I can read on his computer, “There is a stud planted in my leg.” She explains that she had to keep it like that to hold the computer. “Apparently,” says the PA, “there have been numerous statues of Hawking, first with the commutator in the right hand, then the left one.” “We should put the electronic device on his glasses,” interjects the GA, “with long wires behind.”

“The statue will be outside, so we need to make sure that the students won’t sit on its lap,” says one of the physicists jokingly. Another adds, “Stephen would like that, especially if they are girls! [everyone in the room laughs].” Then they talk about the place where the present statue should be kept. One of the physicists offers, "It could be outside," but the sculptor thinks that's not going to be possible, "because it will disintegrate." "Maybe in the corridor," another says. "Well," a physicist adds jokingly, "when students are too frustrated with their unanswered questions, they could ask the statue! [another big laugh]." And "What about his voice?" mentions the other physicist: "We could have his voice working outside! [laugh]." The other one says, "The statue should be wired, then it will electrocute people who sit on it [laugh]." Then the GA jokes that "the statue should be sixty feet tall and could have a restaurant in its head, [laugh]" or "The COSMOS!" adds a physicist [laugh].

"And, yes, what about the head?" says one of the physicists before leaving the room. "After all Hawking is the head, that's where he has his thoughts, *the head should be bigger!*" he says to the sculptor. He adds: "It's true, when I compare the real person and the statue . . ." He pauses and then goes to the door and comes back toward Hawking's wheelchair (he sees Hawking with the statue at the back of the room). He talks about the expression of the statue: "It looks very severe from Hawking's perspective, but when one goes toward the door, it seems more gentle." Then someone says, "I think the GA is blocking the light." The GA moves. Then the PA, talking about the statue says, "Yes, the expression has already changed!" The PA adds, "Yes, it would be fantastic if we could have a sculpture of him with his big smile!" The physicist responds, "No, I think you can't do this, he must look serious." Hawking looks at the statue—I'm confused because I don't understand if they are making fun of him; I can't tell what Hawking thinks. Hawking writes something on his computer. The voice says, *"If I were in zero G, it would be the only way I wouldn't need the wheelchair."* [Everyone laughs.] These few words redirect the conversation. Then someone makes another joke and adds that "the bronze statue could turn in circles around the entire site like a graviton [laugh]"; another interjects, "People could even be inside! [laugh]."

The sculptor mentions that it is time to leave and asks her companion to take a picture of her with Hawking and the statue. I take a picture of them, thinking that it will be useful in analyzing this interaction. She says goodbye to Hawking. The sculptor, her companion, and one of the physicists leave. The electric atmosphere dissipates. The other physicist, who remains in the room, asks Hawking if he wants to go to London next Thursday (apparently an important meeting). "It would be good if you could come along and say something," he adds. Obviously the timing is a problem. They have to be there at nine o'clock in the morning, which means that Hawking needs to get up at four or five o'clock. The graduate assistant in the background mumbles, "Yes, and he could recycle something." Hawking writes that "he has to say something, otherwise he will look like a figurehead." The physicist nods: "Yes, of course, *you have to talk!* It would be great if you could come." Hawking agrees to go. The PA, the GA, Hawking, the statue, one of the physicists, and myself remain. They talk about the statue of Newton at Trinity accompanied by other famous scientists. "*They must be in marble,*" says the physicist. The PA, talking about Newton, says that "he is among a lot of famous ones." The physicist asks the PA if the sculptor is known. This seems important. It's time to go. The door closes. Behind it, Hawking the man and Hawking the statue are left alone. I follow the PA into her room. I ask her about the sculpture. As a response she shows me several statues of Hawking on her computer screen. One day, a picture of this statue will be among them.

AN ETHNOGRAPHIC STUDY OF A STATUE

In the scene described above, I would like to explore the fundamental role of the statue as a site of creation, mediation, articulation, and stabilization of the identity of Hawking the man and HAWKING the genius. The key points of my analysis are the materiality of the statue, its physical or imaginary displacement, and the laughter it provoked. Let's start with the material.

The Material: Plaster and Bronze

The statue is made from a light and malleable material, plaster. The idea is to be able to move it around—to show it to Hawking and then to be able to alter it, in accordance with his comments and those of his assistants. It is a draft, a rough copy, a locus of experimentation where different scenarios are still possible. If there are alterations, they will be the fruit of an agile hand, that of the sculptor. In fact alterations that are wanted, that is, thought of and suggested by the group and implemented by the artist, are acceptable or even desirable. On the other hand, involuntary shocks, caused by clumsiness or time, could disfigure it. It therefore arrives protected, under plastic. Wrapped, it cannot be seen, but people already want to photograph it. Yet not all scenarios are possible. The material also imposes constraints. Remember the comments of the artist when she pointed out the difficulties that she encountered in sculpting the wheelchair. . . . Bronze—luxurious, imposing, solid—has been chosen precisely for the opposite reasons: the final version of the statue will not be able to be moved or altered. It will have to withstand the test of time: immortal, untouched, unchangeable, like the archives extending it, and around which one will simultaneously be able to move.[4] Yet, whether one likes it or not, nothing remains as it is, nothing is ever inalterable, nothing is ever unthinkable, even the most improbable scenarios. Disrespectful students could turn the statue into a bench. In short, qualities are not embedded for eternity in a being, whether a statue or a man.[5]

The Genius

Keeping an eye constantly on the material, let's now follow the statue's movement as it is transported from the corridor to the stage: Hawking's office. It, too, is accompanied by Hawking's graduate assistant. It is undressed; the plastic has just fallen. This first draft is submitted for the man's assessment, as is generally the case with demonstrations, diagrams, articles, conferences, or scripts. Here, everyone is awaiting his approval. The light bothers him. Although uncovered, the statue is actually invisible. The word *black* pronounced by Hawking's synthesizer after much negotiation, is interpreted as a desire to see the statue against a black background. The graduate assistant unhooks the professor's black gown. The man beholds the statue dressed in his cloak; the assistant and the statue behold the man beholding himself.

Hawking is still blinded. This time the word *black* is interpreted as a wish

Figure 8. Photo by Hélène Mialet.

to see the statue placed in front of a blackboard. Someone whispers in my ear that that is where the man always has his photo taken: "Its normal after all, it's always like that, Hawking always has his picture taken with the blackboard behind him" (GA). I interpret this to mean that this is how "he poses for the press"—"when he is the genius." Hawking looks at the statue that has taken his place. The statue has become visible for him: the light doesn't bother him anymore, and for me it has become a locus of comparison.

In the mythology of Einstein and especially in Barthes's analysis of it, the blackboard also plays a fundamental role: "Photographs of Einstein show him standing next to a blackboard covered with mathematical signs of obvious complexity. But cartoons of Einstein, the sign that he has become a legend, show him chalk still in hand, and having just written on an empty blackboard as if without preparation the magic formula of the world."[6] For Barthes, "Mythology shows as awareness of the nature of the various tasks: research proper brings into play clock-work like mechanisms and has its seat in a wholly material organ which is monstrous only by its cybernetic complication; discovery, on the contrary, has a magical essence, it is simple like a basic element, a substance, like the philosopher's stone of hermetists, tar-water for Berkeley, or oxygen for Schelling."[7] What is intriguing here is the strange similarity between what the statue does and what the person "that it represents" does. Neither one nor the other has written these mathematical formulae. Hawking lost the use of his hands long ago; as we have seen, his students and a whole collective to which he is attached do this work for him. He can write no equation, do no calculation by hand, solve no problem without this collective that has disappeared from the scene.[8] All these equations seem to come directly from his head. By taking the place of HAWKING, the statue reminds us that here the man also acts as a statue (see figs. 9 and 10).

Detour toward the Artist

Let's leave Hawking and the statue for a moment, and focus this time on the artist to whom the group addresses its questions. She explains how she went about making the statue. Like the graduate assistant who learns to do his job (i.e., taking care of the various elements—machines and media—that participate in the creation and maintenance of Hawking's agency and identity) by talking to his predecessor, and the journalist who writes about Hawking by talking to and reading other journalists, or the archivist who selects Hawking's writings based on other texts by or about Hawking, the sculptor talks to other artists who had sculpted Hawking. She also works "from life"—that is to say, from photos that enable her to transport Haw-

Figure 9. Hawking the Statue in front of a blackboard. Photo by Hélène Mialet.

Figure 10. Hawking the man in front of a blackboard. Cover of *Une brève histoire du temps*.

king from the office to her workshop.[9] Finally, of course, she will take into account the group's comments (including those of Hawking) and the constraints that, as we have seen, are imposed by the material.

The Computer

Let us revert to Hawking, who has a particular role since he is both subject and actor in the collective production of this statue.[10] He mentions what distinguishes the statue from himself. He questions the need for the bar planted in "his" leg. This is no exact replica; in reality the bar does not exist. To sculpt both the man and the computer without *adding* a support, one would have to use a more solid material. Once again, this is the advantage of bronze. Because it is so fragile, plaster does not allow for a perfect representation. The statue—the first one, they say, to represent him with his wheelchair—is compared in thought to all those that preceded it: those that represented him with the switch in his left hand, and those that represented him with it in his right. One can follow the history of his body materialized in a number of statues: roughly, the transformation of his flesh-and-blood body through his sacred body.[11] For the imitation to be perfect, that is, for the statue to resemble the man as he is today, it would be necessary to represent the new IRS system that he uses to communicate. As the graduate assistant says, "We should put the electronic device on his glasses . . . with long wires behind [them]." Yet, even though the switch has disappeared from his hands, it will not disappear from the hands of the statue. The new system attached to his spectacles is not represented. This is partly because of how the public, and in this case the artist, knows him, and has always known him; partly because the system would be invisible on the statue; and partly because of how Hawking has decided he wants to show himself. The statue will differ from the man as he is today and will not take into account the most recent changes to his body.[12] In its final version, the statue will have a wheelchair and a switch. This is how HAWKING will be immortalized.

The Women

Like a novelist creating his characters, or Woltosz building his machine by taking the place of its users, or Hawking discovering the properties of his objects by throwing them into black holes, the statue is then imagined

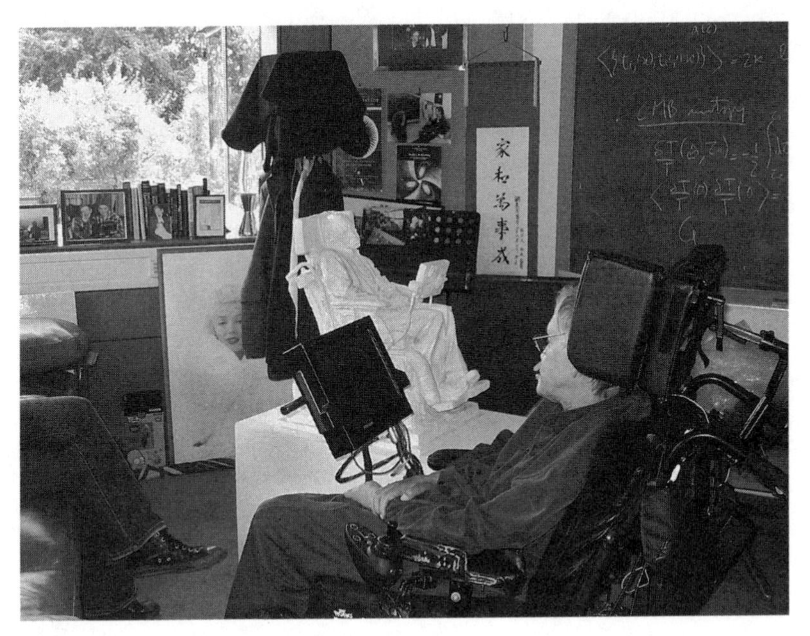

Figure 11. Picture of Marilyn Monroe in the background. Photo by Hélène Mialet.

finished, in bronze, outside. "The statue will be outside, so we need to make sure that the students won't sit on its lap!" says one of the physicists jokingly. Another adds, "Stephen would like that, especially if they're girls!" [laughter]. Although sacred—the incarnation of a person—the statue could become functional, a common object, a bench for students. Yet if the girls sit on his lap, Hawking, the man (!) would like it. The statue and the man are one and the same.[13] The statue, which from sacred has become functional, tells us that the man likes women. That is what makes him human. Faced with skirts, he does not remain cold as stone, or bronze! Passing from the state of dignity of a person to that of an object, the statue tells us that this person is not a statue, that is, an object, and all that with a play on words (see fig. 11).

The Oracle

"The place" then becomes the object of transition. The fact of imagining the future position of the bronze statue redirects the group's attention to that of the plaster statue. After it has fulfilled its role of representation and locus of experimentation, one physicist wonders where it should be stored. Another physicist exclaims, "It could be outside!" but the sculptor says this is impossible "because it would disintegrate." It would indeed be impossible to leave it outside, like the bronze statue, because it would melt. Remember that, unlike the flesh-and-blood body, the sacred body never dies. Therefore, because it is in plaster and fragile, they suggest placing it in the corridor at the entrance to Hawking's office. There it will remain, protected from any mishaps. "So, maybe in the corridor," says someone, and one of the physicists adds, "so that when students are too frustrated with their unanswered questions, they could ask the statue!" (laughter) Once again we have what Koestler characterizes as bisociation—a deliberate confusion between the statue and Hawking, side-by-side imitating each other—that provokes laughter. Once again the statue tells us about the man that here is acting like a statue. Due to the absence, slowness, or brevity of his answers, Hawking has become a sort of oracle. As one of his colleagues said,

> Someone will ask him a question, and it takes a long time for him to construct an answer, and the answer comes, and it has to be simplified, because he can't otherwise express it, and so it comes back rather enigmatic. And so it's hard for young students, I think, particularly, because they see these enigmatic answers, ... and they have somehow to make this work, and it's difficult because they don't quite understand.... And it means he has a kind of oracle role, which is a strange one in a scientific community I think. [...] But you can see why it happens.[14]

As his graduate assistant sums up, "There is a lot of putting words in Stephen's mouth."[15] Recall that Hawking's student confirmed this intuition: "For the first year and a half, each sentence of Stephen's took me about six months to understand. I was six months behind what he was saying, and catching up little bit by little bit. It was hard."[16]

The Voice

Hawking's silence or his enigmatic answers bring to mind his mechanical voice, the synthesizer. "What about his voice?" asks one of the physicists. "We could have his voice working outside!" [laughter again] The other physicist adds jokingly, "The statue should be wired; then it will electrocute people who sit on it." By mentioning his voice, the actors underline the fact that this device has become an integral part of Hawking's identity—for the public but also for him, and partly for that reason. Remember that although his voice has an American accent, Hawking has never wanted to change it, for fear of losing his identity. Yet this voice would seem ridiculous with a statue. One gets used to the idea that a man can be accompanied by a mechanical voice, but not the idea that a statue can be accompanied by one.[17]

The voice also highlights the fact that what characterizes Hawking is that his being is surrounded by a system of electronic devices that are meaningful, paradoxically, because he is human. A man usually has a voice, and even if it is next to him, it is his voice. But a voice placed on a statue that does not have a voice makes this mediation visible and provokes laughter: a statue does not talk. There are limits to the exactitude and conventions of representation. The voice therefore adopts the properties of the person or thing to which it is attached. Yet there is a point of similarity. The voice protects the man. As Albert Robillard, who has Lou Gehrig's disease, says: "I found not having a 'real time' voice equivalent to not having any defense against what was done to my body. . . . I came to visualize having a voice as having the defense of making assertions about myself, making threats or counter-threats, and otherwise carving out and maintaining a space for myself. I could not control what was happening to my body, nor could I control the interactions that largely made up my person."[18] The voice could also protect the statue. Indeed, this system can be seen as a means to preserve the object by electrocuting the unfaithful: a statue turned into an electric chair.

The Head

And then the statue is imagined bigger, by association with other famous statues: "The statue should be sixty feet tall and could have a restaurant in its head!" jokes the graduate assistant, or "The COSMOS!" says a physicist. This new bisociation between the infinite cosmos and the man's finished head is an idea that provokes laughter, although Hawking is very often represented or *drawn* this way, to use Barthes's expression.[19] In passing, we learn that the man does cosmology. It is intriguing that this idea makes them laugh, even though it is also taken literally. In fact there is a shift: laughter clears the way to a more serious level of talk. Everyone's gaze shifts to the head of the statue and the head of the man. Before leaving the office, one of the physicists turns to the sculptor and exclaims, "But yes, and what about the head? After all, Hawking is the head, that's where he has his thoughts, *the head should be bigger*!" The metaphor "the brain in the vat" becomes literal; it is materialized.[20] The metaphor has a body. The statue will not resemble the man—it will incarnate the genius; the bronze statue will have a big head.

We thus see how those who are close to Hawking and constantly do what he cannot do forget their role. Is this not the incarnation of our most beautiful myth, we moderns: the fact of doing science and especially theory by means of nothing (to quote Barthes again) but the "clock-work like mechanism" of a well-made mind, a wholly "material organ which is monstrous only by its cybernetic complication?" This head—this pure mind—that we have seen emerge from journalistic accounts, is literally shaped here as matter and takes on the form of an object—a head that had to be sculpted before being placed on a body; a head that should have contained the cosmos; a head that will be bigger than normal to accentuate the place in which genius resides. This recalls what the director of *The Hawking Paradox* said: "Because there is this fascination with the mind of Hawking and his genius, . . . we did some very close stuff on his eye, which was very powerful. . . . In the real film, I think it's quite a beautiful shot."[21]

The Smile

"The head," this time, is the object of transition. The group starts to compare, not in thought, but by sight: "the real" Hawking, the man, and "the false" Hawking, the statue. Of the former, one sees the face; of the latter, the profile. They are next to each other—they look at each other. People walk around them. By an effect of comparison, the man and the statue are simultaneously endowed with competencies and even exchange their properties.

The statue, immobile and voiceless, goes from serious to soft and vice versa until it comes to life: its expression changes. From Hawking's perspective—he is in the foreground, it is in the background—the statue seems severe. From the door—this time the statue is in the foreground and

Figure 12. Photo by Hélène Mialet.

Hawking in the background—it seems softer. The graduate assistant blocking out the light is told to move. The statue's expression has just changed. Suddenly, it comes to life.[22] In other words, its transformation to being endowed with expressions and emotions happens by an effect of transparency, juxtaposition, and comparison between the man and the statue, the statue and the man: he is serious/it is serious; it appears soft/he is soft; he appears soft/it appears serious, etc.—unless the aim is simply to take Hawking's place to try to see what he is seeing.

This is a strange symmetry, for the processes of attribution through which a statue is endowed with emotions and competencies—which take shape here through comparisons and the play of light—remind us that this is how competencies are attributed to Hawking and how his intentions are read.[23] Like the statue, Professor Hawking is transformed into a mirror of all the projections, as in all human relations, but magnified. We do not know whether this is a Hawking who is suffering, who is bored, or who is thinking.[24] The statue tells us of the man that he is read like a statue. But more than that: it is precisely at the moment when the statue starts to come to life—its expression changes—that someone remembers that what differentiates Hawking, the immobile, voiceless man, from a statue is his expression! The personal assistant, seeing the statue's expression changing, adds, "Yes, it would be fantastic if we could have a sculpture of him with his big smile!" This brings to mind John Gribbin's words: "It is impossible to know the man's inner thoughts, so intimately linked as he is to machines, a set of cold devices enabling him to move, speak and breathe. His face is, if anything, more expressive than most because, aside from his gift for succinct language, it is just about our only window into his mind."[25] And it is in response to his expressions that his intentions are interpreted: "People have been watching when I've had a conversation with him; all they've seen is me questioning and then going 'Right,' 'OK,' and they ask, 'How did you get an answer?' You can see nods and shakes."[26] In other words, the statue's "humanity" reminds us that Hawking's humanity is to be found in his expressions: "the only window into his mind," as Gribbin put it, and especially his smile.

Finally, if the statue is to resemble the man, it is necessary to represent his smile, and that is where, paradoxically, the resemblance must end. In response to the comment of Hawking's assistant on the possibility of sculpting him with his smile, one of the physicists exclaimed, "No, I don't think you can do this, he must look serious." The statue must not resemble the man, nor even seem human; it must incarnate the sacred body—the *gravitas* of the Lucasian Professor of Mathematics; in other words, it must look serious and fit into the tradition of the great, the geniuses, the Newtons and the Darwins. Hawking smiles. But HAWKING will not smile. The statue has to act like a statue and not like the man.

The Zero-Gravity Flight

We then witness a new comment by Hawking, who is often out of phase during the conversation. He notes that if the statue were represented as flying, as he did, the statue would not need a wheelchair. His voice says, "*If I was in Zero Gravity, it would be the only way I wouldn't need a wheelchair!*" He is, of course, referring to the famous zero-gravity flight that he made on April 26, 2007, on a "Vomit Comet," during which he experienced the joys of weightlessness.[27]

These few words steer the conversation in a new direction. Thus, the man/statue flying in the air is a reminder that "the bronze statue could turn in circles around the site like a graviton [laughter]." Another person adds that "people could even be inside [laughter]." What is funny is the idea of the statue flying (but the true Hawking did that) and of Hawking/the statue being compared to a normally invisible graviton, large enough to contain people, like a spaceship. We witness an inversion of proportions.[28] Hawking is seen through what he does, as when his head was imagined containing the cosmos, and we learn about what he does and on what he is working.

Figure 13. The zero-gravity flight. Photo courtesy of NASA/Jim Campbell.

We also recognize the obsession with getting ever closer (as when Hicklin wanted to focus in on his eye), until one is right inside him, to solve the mystery.

The Public Man

Finally, the statue and the man are separated spatially by a new movement. The physicist who a moment earlier compared the man and the statue, has taken a step forward, toward the man this time. The statue remains in the background, in front of the blackboard. The physicist is talking to Hawking and not to the statue. He asks him if he would be interested in going to a conference in London. "It would be good if you could come along and say something," he says. The graduate assistant in the background mumbles, "Yes, and he could recycle something."

Hawking, in what I interpret as a moment of anxiety, ensures, before agreeing, that he will talk: "I have to talk, otherwise I will look like a figurehead." He does not want to resemble a figurehead, a man of straw—a statue! His colleague agrees: "Yes, obviously *you have to talk*! It would be great if you could come." However, as usual, Hawking will be present in person, but everything will have been prepared in advance, down to the last detail. HAWKING will have to be replicated—reproduced in front of another audience. His discourse will be "recycled" by his computer or his assistants, and

Figure 14. Hawking speaking at the White House Millennium Evening Lecture. Courtesy of the William J. Clinton Presidential Library.

will soon be recycled by others, the journalists. He too will be transported from the bureau to the stage. A statue accompanied by a voice—a reminder of the jokes that were flying about earlier!

CONCLUSION

By following the material, metaphorical, and collective construction of this statue (in which Hawking and the statue are both the actors and the subjects), we have seen how, by an effect of comparison between the man and the statue, the qualities of the man, the genius, and the statue emerge. Thanks to the statue, we witness an exchange of properties between it, the man, and the genius, and a constant negotiation between what belongs to the one or the other.

A cascade of events follows. The statue is moved from the wings (the corridor) to the stage (Hawking's office); its attire (its plastic wrapping) is removed, and it is covered with its stage clothes (those that the man wears when he is the Professor: the black gown).[29] It is made visible by being shifted from the window to the blackboard, both literally—the light is no longer a problem—and metaphorically, where it has taken Hawking's place when he is the genius. The man notes what, in the statue, distinguishes it from himself (it has something that he does not have, the plaster bar), and we learn by comparing it with other statues of Hawking what it will represent (something that he no longer has: the switch) and what it will not represent (something that he has today: the blink switch). It is imagined in another place (outside) and in its final form (in bronze), a genius, a sacred being (a person), or as something profane (an object), a bench or an electric chair, with the voice (or not) of the man, with the cosmos in its head, with a big head, or in its final place of residence (in front of Hawking's door). It is compared to the man, no longer by transparency or in thought, but by sight, and has exchanged properties with him (it is endowed with emotions, it has expressions); it is compared to his/its peers, other geniuses/statues, and teaches us what only a statue can (being immobile and serious forever). It goes round in the air for the last time (like the man) or around the department (like a spaceship/graviton), and is set on the ground once again and put aside, as the man comes to the front of the stage. They exchange their properties once more. The man wants to make sure that he will not resemble a statue, for, unlike it, "he will talk." Yet, he will also be transported from the wings to the stage. And as with most stage performances, everything will have been well rehearsed. We thus see how the statue, by moving through different states—from plaster to bronze, shadows to light, invisible to visible, sacred to profane, human to sacred, artifice to nature—makes the man in the room pass through various states (man/genius/statue), by way of a mirror effect.

When the statue takes the place of the man, it reminds us that he is a statue when he is the genius. When the man compares himself to his double, he highlights that which distinguishes him from it and that which will distinguish it from him. When the statue becomes an object, it says of the man that he is a subject. When it is placed next to his office, it says of the man that he acts as it does. When they are compared visually, the statue comes alive and tells us that what distinguishes the man from the statue are his expressions and his smile. It simultaneously tells us what will distinguish the statue from the man: the statue must neither smile nor seem human; it must resemble a statue. Then the man expresses his fear of resembling the statue, even though he will act like one. It is in moments of closeness (in thought, words—play on words—and appearance) that the exchange of properties takes place, and that we learn what distinguishes the man from a statue, or what makes a statue of him, and what only a statue can do.

Thanks to the statue, as a mirror or a negative example, we have discovered the qualities of the man and what is peculiar to him: he has a body of flesh—with no bar planted in his leg—and an infrared ray attached to his spectacles; he likes women; he does cosmology; he is a head; he smiles; he has a mechanical voice; he has flown in the air. We have also learned that the man shares certain qualities with the statue or that the man acts as a statue, in the sense of having intentions or competencies attributed to him, like that of "genius," whereas he is not at the origin of the action: "when *he writes* mathematical equations," "when *he answers* the students' questions," "when *he gives a lecture,*" in other words, when he poses for the press, when he teaches, when he performs. It is the students who write and solve equations, and who lend him their words. It is the assistants and the computer that preprogram his performance. Here, the original is made: like a statue, the original is a fake—Hawking has done little or nothing. The original is a mirror: in him one reads one's own words; the original is a copy: Hawking's discourse has been, is, and will be recycled; the original is a replica: the performer has been replicated. Finally, we have discovered certain qualities that seem to belong only to the final statue: it will have a switch in its hand, it will be in bronze, it will have a big head, and it will be serious. The copy is the original. It is the genius. As for Hawking, as a member of the group, even if he spoke little, he definitively altered the interaction by causing the statue to move, by criticizing or imagining his appearance differently, by provoking laughter, by allowing for a visual comparison, by being afraid of resembling it.

Hence, without the plaster statue, there would have been no laughter, jokes, and comparisons between the true and the false, no photos, sculptor, or ethnologist; only a few physicists and assistants going about their business. With the statue, and without Hawking, there would probably have been laughter, jokes, comparisons between the true and the false, photos, the sculptor, the ethnologist, the physicists and the assistants, but the final statue would have been different.

Yet, does this mean that Hawking the genius is only the fruit of random processes of attribution? This is not what the actors seem to say. Indeed, they incessantly alternate between different levels of speech that sometimes seem to mock the man—HAWKING is a statue (an oracle)—and sometimes to venerate him—Hawking is his mind, the statue ought to have a big head! As I left, I couldn't help but wonder: Do physicists believe in their myth?[30]

The social sciences have for a long time made subjects into silent agents—statues—that are put into action and into words by others, society, culture, habitus. Or, on the contrary, they have made them all-powerful by overlooking nonhumans. Rethinking the role of a statue has enabled us here to rethink the role of the subject, and not any random subject, but one that embodies the mythical figure of the isolated genius capable of grasping the ultimate laws of the universe on the sole basis of his reasoning—the knowing subject—the thinker;[31] now a man, now a genius, now a person, now a statue, now an actor, now an actant.

Conclusion—A Recurring Question

From Exemplum to Cipher

Our traditional understanding of scientific knowledge is based on notions of the lone genius, the pure mind, and the rational actor. In a sense, this is what makes us "modern"; we have no need of a body, we don't need to be situated, we don't need colleagues, we don't need institutions: to be able to think, we just need to have a good mind. This is the conception of knowledge against which the field of science and technology studies has constituted itself. Indeed, such studies have shown that scientific knowledge is not (simply) the product of rational individual minds, but is an eminently social and material process. In this regard, science is not different from other forms of practice: it is a socially, materially, historically situated form of life. This book has been an attempt to test this hypothesis through the case study of a man who is considered the most singular, idiosyncratic, rational, and bodiless of all: Stephen Hawking. It is also an attempt to conceptualize the knowing subject anew.

Stephen Hawking, I have argued, is not different from other scientists because he is disabled; rather, it is because he is disabled that he offers us the possibility of seeing how scientists work. In other words, Hawking enables us to see collective and material practices normally hidden from view, and without which scientists would not be able to perform. In this sense, he is an exemplum—a type. If there are differences, they are not in essence, but in degree.

More generally, I have shown that the most extreme individuality—a man reduced to a minimum of action, a man who can't move and today only communicates with his eyes—makes visible a collective through which his intelligence is distributed, or more precisely composed and recomposed.[1] In other words, you thought, and I thought, especially when I started this project, that I was studying "a man"—"an individual"—when what I was studying was a large and complex organization, in the same way that Hutchins studied a cockpit or a ship, or the Goodwins an airport,[2] with the exception that *he is an individual*. Or to put it differently, the individual, I argue, *is* a collective.

If ethnographic studies of everyday practice, of situated cognition, language, and interaction, were an important source of inspiration for this book, actor-network theory was also very influential in helping me understand the constitution of Hawking-the-great-researcher. In a certain sense, I have become more royalist than the king, for I have argued that Hawking materializes, *gives flesh* to, the actor-network; indeed, he is *literally* an actor-network. If we go back to the metaphor used by Bruno Latour when he says that Pasteur-the-great-researcher is the product of a collective constituted of humans and nonhumans, or, to quote Callon and Law, that "Pasteur-the-great-researcher [. . .] does not exist outside this network, which strictly speaking, constitutes his body and mind,"[3] we can say that in Hawking's case the metaphor is materialized. Indeed, not only are Hawking's intellectual competencies incorporated, materialized, and distributed in the networks of humans and machines I have been at pains to describe, but so too are his identity and his flesh-and-blood body. In this regard, he is not just an actor-network, he is the epitomization of a cyborg, to use Haraway's terminology.[4] Thus, the one who is supposed to be the most singular, individual, rational, and bodiless of all is not only incorporated, materialized, distributed, exteriorized, and mediated *like* everyone else, but even *more* so.

If HAWKING and Hawking are a corporation, an organization, a collective, an actor-network, and a cyborg, does this mean that the subject—the individual—is then lost? No. This is my second move. Though one might conclude that because Hawking is a cyborg, a totally distributed postmodern subject, an actor-network composed of humans and nonhumans that constitute his mind, his identity, and his body, we necessarily lose sight of the subject, I argue, on the contrary, that it is precisely because of this material-collective process that we can grasp his singularity. This is not a contradiction. Rather, by showing the (re)distribution of competencies to nonhumans and humans as an antidote to the image of the ethereal, rational subject, I have tried to reframe questions of singularity and individuality. Indeed, they are central to this project. Thus, it is not because Hawking is outside the social and material world (i.e., the brain-in-the-vat) that he is the most singular; rather, it is because he is the most materialized, the most collectivized, the most distributed, and the most connected that he is the most singular. He is what I call a distributed-centered subject.[5]

The notion of the distributed-centered subject derives from my first book, *L'Entreprise créatrice*. Based on an ethnographic study conducted in

1992 in a major French oil company (Elf Aquitaine, now Total), this book presents a picture of how a research scientist (François Montel) working for a large multinational corporation distinguished himself through the creative skills he applied in formulating working instruments, organizational structures, and human relations. I argued that the more this actor was linked up with his institution, his objects of research, his co-workers, and so on, the more potential he had to become inventive; and the more inventive he became, the more he seemingly distinguished himself as an inventor, as a kind of genius who existed beyond social, material, and cultural constraints. It is, I argued, by simultaneously "becoming" his object of research, personifying his institution, and empathizing with the needs and interests of his colleagues that he was able to invent. By placing himself at the intersection of different research fields, he was able to translate problems from one field to another and thus constantly feed his inventive mechanism and his status. By consolidating his group from the inside, that is, by making it participate in the creation and by delegating to it his know-how, he extended his recognition. By locating the computer programs he had invented in different phases of interpretation of oil fluid, he spread out his action. By placing doctoral students—whom he would eventually bring back to his lab—in well-known research laboratories, he gave himself the opportunity to fuel his theories with new information and also to spread his field of influence. In short, it was because he positioned himself, as the semioticians would say, in several places in order to occupy different spaces, that he was able to feed his know-how, maintain himself at the center of the network that he created, and enhance his recognition. He could be distinguished from this network precisely because he moved within it in a specific way, mixing with his research object, and translating and linking up things that were previously unconnected. Indeed, in the case of Montel, the presence of the inventor's body was indispensable to understanding how certain properties were exchanged and how an innovative process functioned.[6]

Stephen Hawking, similarly, seems more like the inventor working in a firm with a thousand other persons than the solitary scientist he is usually described to be. Moreover, he is even more distributed than François Montel insofar as his inability to make his own body work has allowed numerous operations, normally embodied in a single, able person, to become visible because externalized and incorporated in other bodies. With Montel, I followed the movements of humans and narratives. In the case of Hawking, I highlight other mechanisms of singularization.[7] For example, Hawking's singularity is constructed by the different media at work; one of them, the press, constantly recycles tropes having to do with his American accent, the coincidence of his birthday with the year of Galileo's death, the role of his wife in his survival, and so on. At the same time, his singularity also appears in the nodes and interstices of the networks of humans and machines to which he is attached: he resists, he doesn't want to change his programs, he doesn't want to change his voice, he doesn't want to change his Web page, he runs over the toes of those he doesn't like, he plays with his yeses or nos, he plays with his disability, or he refuses to delegate.[8]

The idea of a decentered subject has become fashionable today: we are all, in one way or another, constantly connected, extended, wired, and dispersed in and through technology. One wonders where the individual, the person, the human, and the body are, or, alternatively, where they stop. These are the kinds of questions I have tried to reconceptualize by focusing on a man who is, as I have shown, permanently attached to assemblages of machines, devices, and collectivities of people. Indeed, if Hawking is a corporation, he also has an extended body or, more precisely, a multiplicity of extended bodies, of which he is at once an element and a product—hence the double meaning of the title of this book *Hawking Incorporated*.[9]

One of the challenges of my study has been to make these bodies visible and to show how they disappear and reappear. This has proven especially difficult, for even when they are visible, we tend not to see them. Indeed, it is hard to know where exactly Hawking is. This is why in the first chapter we followed the reproduction of HAWKING from one point of the planet to another, tracing the transformation of the slightest movement of Hawking's eyebrow to HAWKING the genius talking to a crowd of enthusiastic fans. A world appeared, consisting of the graduate assistant, the personal assistant, the computer, and the synthesizer, all of which made themselves visible or disappeared, depending on the situation. In the second chapter we continued the work of deconstructing (or reconstructing) the genius, this time by describing the network of competencies that enabled the man to produce theories. Another part of his extended body appeared: students/computers/department/university. In the third chapter we opened a different black box: that of his mind. Here we saw the operations of the students and the diagrams thanks to whom and through which Hawking can move and travel in infinite space. In the fourth chapter we looked at the press and the journalists, and followed how, in concert with the scientist, the image of the genius that we all know is constructed. In the fifth chapter we questioned/interviewed HAWKING, losing sight of the *genius* in the ramifications of the network constituting him, only to find the *man* reappear again in the act of constructing his being/persona. In the sixth chapter we turned to the archivists working on storing Hawking's memory and participated in the construction of the identity of this self over time (or its inscription in time: its immortalization). In the final chapter we studied the production of the statue of Hawking to the point of being lost in a hall of mirrors, bouncing between copies and originals.

Thus, each chapter of this book has focused on describing the operation and coordination of different elements or media (the mediations) that create the presence, the agency, the identity, and the competencies of Hawking. Moreover, each chapter followed a process that resulted in a product: a talk (chap. 1), a scientific article (chap. 2), a thought experiment/a diagram (chap. 3), a documentary film (chap. 4), an interview (chap. 5), an archive (chap. 6), a statue (chap. 7). Finally, each product was attached to a form of action: *He speaks* (chap. 1), *He writes* (chap. 2), *He thinks* (chap. 3), *He is* (chap. 4), *He converses* (chap. 5), *He will be* or *He was* (chap. 6), *He incarnates* or *He is* (chap. 7). In other words, when one says, "He gave a talk," this statement stands for and is the final product of more than fifty pages of descriptions of a complex network at work. This is true of every statement, every product, and thus of every chapter of this book.

Instead of always moving among three paths—the genius as a pure mind, the genius as a pure social construction, or the genius as a product of a network composed of humans and nonhumans—I have tried to follow another trajectory and propose a new methodology. We are close to the man, and then we are far from him, and then close to him again. We track movement as a camera does. Sometimes the man appears, sometimes he disappears; sometimes the collective or a part of the collective he is connected to appears, sometimes it disappears; sometimes the genius appears, sometimes he disappears. This is why I didn't try to understand who Hawking was, but rather *where* he was. More specifically, I have tried to show the different layers that constitute HAWKING.[10] In the face-to-face interaction (without the mediation of the machine), there is little that is intelligible; when we add the computer, the assistants, the students, the colleagues, the diagrams, the media, the analyst, the archivists, the artist, and the statue, we have HAWKING, but in the interstices we can still detect a man fighting to exist. Slowly, the genius, the knowing subject, the person, the individuality, the human being appears like a pack of fanned-out cards. Sometimes they are superimposed on one another; sometimes they are not. Depending on the distance I have allowed myself to take, sometimes we see the genius, and sometimes we don't.[11] Depending on the materiality we deal with—the practices, daily events and activities, or digitized and printed words and images—sometimes we see the genius, and sometimes we don't.[12]

In a certain sense, this approach questions the premise that proximity and direct access are superior to—truer than—texts. Indeed, being close is not always an advantage; proximity is not a guarantee that we can grasp the genius or even the person. In the present case, it was the other way around (see chap. 5).[13] Yet, at the same time, the ethnographic approach was indispensable, for it allowed me to reintroduce the flesh, affect, human and nonhuman prostheses (machines, devices, language), practices, and narratives of which the man, but also the thinker, is made (to echo the role of Ryle's "What Is 'Le Penseur' Doing?" in the creation of Geertz's "thick description"). Accordingly, it is by keeping an eye on the different materialities—practices, oral and written discourses—and by showing the work of collectivization, materialization, and distribution (composition and recomposition) of competencies, on one hand, and the work of singularization, on the other, that I was able to see in action the dual process of mediation (or hybridization) and purification so specific to modernity.[14] Indeed, is this not what we are all doing, we moderns, administrators, and technicians (chap. 1), scientists (chaps. 2 and 3), journalists (chap. 4), human/social scientists (chap. 5), archivists (chap. 6), and artists (chap. 7)? By doing, performing, glossing, and gossiping, we are simultaneously collectivizing, materializing, and distributing Hawking's competencies, identity, and body. At the same time, we are purifying and singularizing him. That is, we are making the genius (see the assistants working for months but disappearing during the performance; the students doing the calculations or drawing the diagrams but attributing the results of their efforts to Hawking; the journalists reconstructing the genius from scratch; the human/social scientist writing a book to deconstruct the genius but saving the subject; the archivists meticulously separating the scientist from his context; the scientist teasing Hawking and simultaneously asking the artist to make a statue of him with a big head because he is a genius).[15] We thus see the moderns—that is, the non-moderns—in the process of constructing their myth: the myth of the moderns, the brain-in-the-vat.

Though Hawking embodies the representation of the genius and, by extension, the traditional conception of what science is supposed to be—a world made of theories produced or "found" theoretically, a world made by neutral, objective scientists who live in the pure and unadulterated world of the mind—he is, in fact, materialized and distributed in a series of overlapping and interconnecting collectivities.[16] This oscillation—this liminality—puts into question many of the fundamental distinctions upon which our scientific rationality is based: the distinction between we "moderns" with our rational minds and the "savage" others, the bricoleurs; the distinction in our society between those who think theoretically and those who work with their hands; the distinction in our scientific laboratories between the geniuses and the assistants; the distinction between the humans and the nonhumans. The assistants are supposed to execute what the chief or the master has thought of beforehand. We saw with Hawking that far from only executing, they do most of the work; they complete, classify, attribute meaning, translate, and perform.

Humans, we are told, have subjectivity—a self, a consciousness, a rationality; nonhumans, on the other hand, are passive, neutral, and inert. However, Hawking's case demonstrates that the machine, for example, does much more than transcribe his thoughts; rather, it makes him write or writes with him: it completes his sentences, it gives him a voice, it gives him the possibility of escaping questioners expecting a yes or a no, it disrupts, and it also makes it possible for people to identify with him. The voice here plays a role not opposed to writing—to logos—for the written precedes (is) the spoken. The voice too is indispensable to the fabrication of his identity—without it he can't perform, and to a certain extent the voice becomes—is—him, as, for example, when it performs without him in the office of his assistant.

However, Hawking is not so special that one cannot generalize from his case. On the contrary, I have shown that he makes visible what is normally invisible or not thought of. Indeed, he makes visible how pop stars and politicians manage their images (chap. 1),[17] how physicists work with their students (chap. 2), how physicists think geometrically versus analytically (chap. 3), how the media recycles stories (chap. 4), how we communicate (chap. 5), how the digital age redefines the process of recording (chap. 6), how objects make us do things (chap. 7), and so on. In this sense, Hawking's case is not limited to science but reveals aspects of other forms of practice. For example, this case helps us understand not only what it means to "do" abstraction, but also what it means to be a manager. In the same way, the notion of the distributed-centered subject could be applicable to fields such as art, industry, or politics.[18]

Stephen Hawking's flesh-and-blood body has, of course, a particular place in holding this nexus of complex networks composed of machines and humans together. However, elements that constitute his being are already beginning to be distributed.[19] His conferences, papers, scripts, and movies are being reorganized, classified, recorded, preserved, and put together in what will be the Hawking Archive. The statue will stand on the other end of the Betty and Gordon Moore Library, where the archive will be housed. His old voice is already in the science museum in London, his wheelchairs and the old computers might be stored in the Whipple Museum in Cambridge, and so on—these are all permanent traces of his presence. Yet, one more time, one can't help but wonder: Where are you, then, Mister Hawking?

Epilogue

Yesterday, as I was wandering through the streets of Cambridge, Hawking appeared to me, as if to thumb his nose or wink at the strange question that has run through this book. But of course he is here, situated, in flesh and bone, in his little car, accompanied by his nurse, traveling up and down the streets of Cambridge just as he travels along the lines of his diagrams. He is here—when he is not out there in the universe—wheeling along the curbs that now match the shape of his body: extended. He has probably just left the Department of Applied Mathematics and Theoretical Physics, leaving behind him his busy world. They say he has to go to Berkeley soon. He's expected over there, in California; they're preparing for his visit and have promised that all the students will have read *A Brief History of Time*. The press has been contacted. A journalist calls me. In Cambridge, behind the department, through a shaded window of the Betty and Gordon Moore Library, appears the silhouette of a young archivist busily going up and down stairs, opening boxes, classifying and "recontextualizing" documents and papers. Early tomorrow morning they'll be sent to the University Library in Cambridge. In a few months copies of "the originals"—which have been left in the hands of historians—will be sent back again to the Betty and Gordon Moore Library, where they will ornament the first floor for the curious eyes of the public. On the second floor of the London Science Museum, the "old voice" of Hawking, classified amongst the latest technological novelties, is shrouded in darkness. The museum closes. A few miles away, Eve Shepherd redoes Hawking's head for the third time. At the same time, in Cambridge, Hawking goes home. But he won't be staying there for long. He's promised to leave planet earth soon to explore outer space, this time with the help not of diagrams but of the powerful engine of a spaceship being built by Richard Branson's company, Virgin Galactic. Talk about abstraction!

NOTES

Introduction

1. John Locke, *An Essay Concerning Human Understanding* (London: Oxford University Press, 1988), 160.
2. Ibid., 162.
3. Ibid., 163–64.
4. The doctor gave him two years to live after the first symptoms appeared; he has now been living with the disease for more than forty years.
5. Jeremy Hornsby and Ian Ridpath, "Mind over Matter," *Telegraph Sunday Magazine*, October 28, 1979, 44.
6. John Boslough, "Stephen Hawking Probes the Heart of Creation," *Reader's Digest*, February 1984, 39.
7. Leon Jarroff, "Roaming the Cosmos," *Time*, February 8, 1998, 34.
8. Jerry Adler, Gerald C. Lubenow, and Maggie Malone, "Reading God's Mind," *Newsweek*, June 13, 1988, 36.
9. Michael White and John Gribbin, *Stephen Hawking: A Life in Science* (New York: Dutton, 1992), 135.
10. Simon Schaffer shows how the term *genius* was applied to natural philosophers in the late eighteenth century. The Romantic genius, he explains, possessed extraordinary, even mystical, power that enabled the natural philosopher to divine nature and discover its secrets. See Simon Schaffer, "Genius in Romantic Natural Philosophy," in *Romanticism and the Sciences*, ed. Andrew Cunningham and Nick Jardine (Cambridge: Cambridge University Press, 1990), 82–98. Steven Shapin has written an important essay on the theme of solitude, "The Mind Is Its Own Place: Science and Solitude in Seventeenth-Century England," *Science in Context* 4, no. 1 (1991): 191–218.
11. Shapin, "The Mind Is Its Own Place," 210. For an inquiry into the changing notions of science as a vocation, see also Steven Shapin, *The Scientific Life: A Moral History of a Late Modern Vocation* (Chicago: University of Chicago Press, 2008).
12. See for example, Steven Shapin, *A Social History of Truth: Civility and Science in Seventeenth-Century England* (Chicago: University of Chicago Press, 1994); Otto

Sibum, "Reworking the Mechanical Equivalent of Heat," *Studies in History and Philosophy of Science* 26, no. 1 (1995): 73–106; Bruno Latour and Steve Woolgar, *Laboratory Life: The Social Construction of Scientific Facts* (Beverly Hills, CA: Sage, 1979); Simon Schaffer, "Making Up Discovery," in *Dimensions of Creativity*, ed. Margaret Boden (Cambridge, MA: MIT Press, 1994), 13–55.

13. Karl Popper, *The Logic of Scientific Discovery* (London: Hutchinson, 1972), and *Conjectures and Refutations: The Growth of Scientific Knowledge* (London: Routledge and Kegan Paul, 1969).
14. See, for example, Latour and Woolgar, *Laboratory Life*; Michael Lynch, *Scientific Practice and Ordinary Action: Ethnomethodology and Social Studies of Science* (Cambridge: Cambridge University Press, 1993), and *Art and Artifact in Laboratory Science: A Study of Shop Work and Shop Talk in a Research Laboratory* (London: Routledge and Kegan Paul, 1985); Karin Knorr Cetina, *Epistemic Cultures: How the Sciences Make Knowledge* (Cambridge, MA: Harvard University Press, 1999); Andrew Pickering, *The Mangle of Practice: Time, Agency and Science* (Chicago: University of Chicago Press, 1995).
15. Thus, for example, there is no radical difference between a physicist and a baker as far as the cognitive operations mobilized are concerned. The difference derives mainly from the objects they handle.
16. For Bruno Latour, "An actor in ANT is a semiotic definition—an actant—that is, something that acts or to which activity is granted by others. It implies no special motivation of human individual actors, nor of human[s] in general. An actant can literally be anything, provided it is granted to be the source of an action. . . . If a criticism can be leveled at ANT, it is on the contrary its complete indifference for providing a model of human competence." "An Actor-Network Theory: A Few Clarifications," paper presented at the Center for Social Theory and Technology workshop, Keele University, UK, May 2, 1997, 4.
17. The principle of generalized symmetry has been extensively discussed and debated. See, for example, the article by Harry Collins and Steve Yearley, "Epistemological Chicken"; Michel Callon and Bruno Latour, "Don't Throw the Baby Out with the Bath School! A Reply to Collins and Yearley," in *Science as Practice and Culture*, ed. A. Pickering (Chicago: University of Chicago Press, 1992); and also the controversy between David Bloor and Bruno Latour in the following three articles: David Bloor, "Anti-Latour"; Bruno Latour, "For David Bloor and Beyond: A Reply to David Bloor's 'Anti-Latour'"; and David Bloor, "Reply to Bruno Latour"; all published in *Studies in History and Philosophy of Science* 30, no. 1 (1999): 81–112, 113–29, and 131–36. See also Pierre Bourdieu, *Science de la science et réflexivité* (Paris: Editions Raisons d'Agir, Collection "Cours et Travaux," 2001); Hélène Mialet, "The 'Righteous Wrath' of Pierre Bourdieu," *Social Studies of Science* 33, no. 4 (2003): 613–21.
18. See, for example, Harry Collins, *Changing Order: Replication and Induction in Scientific Practice*, 2nd ed. (Chicago: University of Chicago Press, 1992); and Barry Barnes, *Scientific Knowledge and Sociological Theory* (London: Routledge and Kegan Paul, 1974).
19. I take this quotation from a draft paper eventually published by Michel Callon and

John Law as "After the Individual in Society: Lessons on Collectivity from Science, Technology and Society," *Canadian Journal of Sociology* 22, no. 2 (Spring 1950): 169. The published version is worded differently; I prefer the draft version.

20. Donna Haraway, *Simians, Cyborgs and Women: The Reinvention of Nature* (London: Free Association Books, 1991).
21. Yes, we say (and he says), because to do theoretical physics, one does not need to have a body, just a good mind.
22. For many, being or knowing is synonymous with doing (see, for example, Annemarie Mol, Andrew Pickering, John Law, Michel Callon, Bruno Latour, Antoine Hennion, Judith Butler, Karen Barad, and Donna Haraway, to cite a few). Thus, understanding what *doing* is becomes crucial.
23. Indeed, if in the rationalist tradition the driving force of knowledge is inscribed in the subject, this subject is devoid of subjectivity: a subject that is transparent for Descartes, desingularized by Kant, or evacuated in Popper. See for example, René Descartes, *Méditations métaphysiques* (Paris: Flammarion, 1979), and *Discours de la méthode—pour bien conduire sa raison et chercher la vérité dans les sciences, (plus) la dioptrique-les météores et la géométrie qui sont des essais de cette méthode* (Paris: Fayard, 1987); Immanuel Kant, *Critique of Pure Reason*; trans. and ed. Paul Guyer and Allen Wood (Cambridge: Cambridge University Press, 1999); Karl Popper, *Objective Knowledge: An Evolutionary Approach* (Oxford: Clarendon Press, 1972) see also the analysis by Evelyn Fox Keller, "The Paradox of Scientific Subjectivity," in *Rethinking Objectivity*, ed. Allan Megill (Durham, NC: Duke University Press, 1994): 313–31; Henry Michel, *Généalogie de la psychanalyse* (Paris: Presses Universitaires de France, 1985), 61: "Une subjectivité privée de sa dimension d'intériorité radicale, réduite à un voir, à une condition de l'objectivité et de la représentation." See also Allan Megill, "Introduction: Four Senses of Objectivity," in *Rethinking Objectivity*, ed. Allan Megill (Durham, NC: Duke University Press, 1994), 1–20, at 10: "Insofar as one stresses the universality of the categories—their sharedness by all rational beings, one will see Kant as a theorist of absolute objectivity, an objectivity stripped of everything personal and idiosyncratic."
24. I deliberately play with the singularity of *expert, intellectual, scientist*, and *human* versus the plurality of *laypersons, workers, technicians*, and *nonhumans*.
25. In this sense I would argue that philosophy and sociology, for opposing reasons, deny the necessity of the situated body of the scientist. Some readers may be surprised to hear that sociologists of science "deny the necessity of the situated body of the scientist," when one of the principal contributions of the social and cultural studies of science has been the work of "reincorporating" scientific intelligence into its environment. For example, see Christopher Lawrence and Steven Shapin, eds., *Science Incarnate: Historical Embodiments of Natural Knowledge* (Chicago: University of Chicago Press, 1998). Nevertheless, I would like to underline that insofar as scientific knowledge has been reincorporated into the social world (as an antidote to rationalist/individualist conceptions of science, for example), the "situated and singular" character of a body endowed with its own idiosyncratic competencies tends either to be dissolved into a "collectivity," or to be "black-boxed" as, say,

"tacit knowledge." On this point, see Hélène Mialet, "Reincarnating the Knowing Subject: Scientific Rationality and the Situated Body," *Qui Parle?* 18, no. 1 (2009): 53–73. On Haraway's notion of "situated knowledges," see Charis Thompson, *Making Parents: The Ontological Choreography of Reproductive Technologies* (Cambridge, MA: MIT Press, 2005), chap. 2, where she notices that "partiality, embodiment, and 'mobile positioning' share sensibilities with recent feminist epistemologies based on motion rather than position, . . . despite the somewhat static sound of Haraway's word *situated*" (49). Concerning the critique of tacit knowledge as a "black box," see the way the historian of science Otto Sibum deals with the problem of reintroducing the power of the productive body, in "Les gestes de la mesure: Joule, les pratiques de la brasserie et la science," *Annales Histoire, Sciences Sociales*, nos. 4–5 (July–October 1998): 745–74. On the distinction between the body and embodiment, see Katherine Hayles, *How We Became Posthuman: Virtual Bodies in Cybernetics, Literature, and Informatics* (Chicago: University of Chicago Press, 1999).

26. Nevertheless, it could result in new ways to write a biography.
27. John Boslough, *Stephen Hawking's Universe: Beyond the Black Holes* (London: Fontana, 1984); Kitty Ferguson, *Stephen Hawking: Quest for a Theory of Everything* (New York: Bantam Books, 1992); Jane Hawking, *Music to Move the Stars: A Life with Stephen* (London: Macmillan, 1999); Joseph P. McEvoy and Oscar Zarate, *Introducing Stephen Hawking* (New York: Totem Books, 1995); Kip Thorne, *Black Holes and Time Warps: Einstein's Outrageous Legacy* (New York: W. W. Norton, 1994); Michael White, and John Gribbin, *Stephen Hawking: A Life in Science* (New York: Dutton, 1992).
28. I will use the style HAWKING for his name, as shown here, when I refer to Hawking the genius.
29. See Ernst Kantorowicz, *The King's Two Bodies: A Study in Medieval Political Theology*. Princeton, NJ: Princeton University Press, 1957.

Chapter One

1. Among the assistants, I also include the designer of the assistive technologies Hawking uses.
2. At the time I conducted my ethnographic research for this chapter (1999–2002), Hawking was using his fingers. He now uses his eyes. As we will see, the configuration of devices he uses changes in accordance with the changing needs of his body.
3. Stephen Hawking, "Prof. Stephen Hawking's Disability Advice," online at http://www.hawking.org.uk/index.php/disability; see also Stephen Hawking, "Prof. Stephen Hawking's Computer Communication System," online at http://www.hawking.org.uk/index.php/disability/thecomputer.
4. Hawking, "Hawking's Disability Advice."
5. Hawking, "Hawking's Computer Communication System." Woltosz designed the software, while David Mason (Hawking's second wife's ex-husband) is responsible for the hardware. It was through Elaine that Woltosz met David Mason. Mason taught Hawking how to use the device; he would subsequently become the Words Plus representative in England.

6. The programs I describe here are the precursors of today's technology. In this regard, they will seem familiar to the reader. We are—through e-mail, Facebook, Twitter, texting, and so forth—increasingly reliant on writing. Stephen Hawking was using these technologies from their inception. Indeed, he is entirely dependent on them and on the technology of writing itself. Hawking's example thus become an interesting window on our current socio-technological and literary practices.
7. Walt Woltosz, interview by the author, December 22, 2000. Each time Woltosz or the GA tried to demonstrate the program, each one clicked on the wrong letter.
8. Software developers used the work of David Beukelman and others in the augmentative and alternative communication fields on the phrases and words most often used in everyday conversation to help determine what should be included in their software. See David Colker, "Giving a Voice to the Voiceless: Technology—Software Enables Severely Disabled People to Communicate By the Touch of a Button," *Los Angeles Times*, May 13, 1997, 8.
9. With EZ Keys, the word-prediction database can contain up to five thousand words and can be easily modified to include new vocabulary. The Word Completion feature includes words used in general (that is, by anyone in everyday conversation), while Word Prediction includes words used by Hawking in particular.
10. Walt Woltosz, interview by the author, December 22, 2000.
11. Words Plus brochure: *Catalog of Products, Alternative and Augmentative Communication, New Products!* (2000), 5.
12. Walt Woltosz, interview by the author, November 27, 2002.
13. Ibid.
14. For example, in the menu for "Conversation," sentences like, "Please, let me finish saying what I'm trying to say!" will be available (Woltosz, interview by the author, 2000), while in the one for "Insults," one can, for example, choose between "What a slob" or "If brains were gunpowder, you wouldn't have enough to blow your nose." See Colker, "Giving a Voice to the Voiceless." The fact that insults are part of the menu tells us something about our expectations of what constitutes an individual (i.e., a capacity to resist). On this point, see Ingunn Moser and John Law, "Good Passages, Bad Passages," in *Actor Network Theory and After*, ed. John Law and John Hassard (Oxford: Blackwell, 1999), 196–219.
15. Woltosz recounts that he has been struggling with the problem of word prediction for years: "We have not been able to think of something for years that would make a significant improvement. We make a little improvement here and there, . . . but the big improvements were abbreviations, word prediction, instant phrases, access to standard software—they have been around for at least five or ten years. So, it is very difficult for us to figure out [how] to make it better. Perhaps a more intelligent word prediction, . . . something that would make the computer sense what it is that you are talking about. Something that would even . . . this is a far-out concept . . . but now voice recognition is getting pretty good. Now, voice recognition wouldn't do Hawking any good because he has no voice; . . . but if the computer could listen to his conversation partner and sense what the conversation is about, then it might do a more intelligent prediction. If I'm sitting here working on a paper about black

holes and radiation and quasars and all that, and someone walks into the room and says, 'What would you like to have for dinner?' If my computer could listen to that person and trigger on the word *dinner*, all right, we are going to talk about food, then it might be smart enough to say, 'Switch my vocabulary to a food vocabulary instead of my quasars and black holes oriented vocabulary for what I'm going to say.' That would be a fairly advanced concept. [. . .] The person can change, but to have it be automatic would be neat" (Woltosz, interview by the author, December 22, 2000).

16. Ibid.
17. Ibid. Woltosz says that Hawking could be faster with EZ Keys. The number of words per minute is not standardized. Hawking says on his Web page, "Fifteen words a minute." John Gribbin in his book says, "Hawking found he could manage about ten words a minute. 'It was a bit slow,' he has said, 'but then I think slowly, so it suited me quite well.'" See Michael White and John Gribbin, *Stephen Hawking: A Life in Science* (New York: Dutton, 1992), 236. This last quotation is reused in Kitty Ferguson, *Stephen Hawking: Quest for a Theory of Everything* (New York: Bantam Books, 1992). In chapter 4 we examine how quotations are reused and recontextualized. This process of repetition plays a crucial role in the emergence and stabilization of what makes Hawking Hawking—that is, his physical, intellectual, and personal qualities.
18. Abbreviation exists only on EZ keys. As Woltosz mentions, "I don't know if Stephen uses it in EZ Keys, but I presume he does. It has been part of the EZ Keys line (originally called WSKE-Words Plus Scanning Keyboard Emulator) since the mid-80s" (Woltosz, interview by the author, November 27, 2002).
19. Woltosz, interview by the author, December 22, 2000.
20. Brochure, Words Plus, for the year 2000, 5.
21. Woltosz, interview by the author, November 27, 2002.
22. Ibid.
23. Or again, commenting on EZ Keys efficiency: "Yes, the visual presentation, the additional features, the pre-organization, are all far far more efficient, but *he is so stubborn!*" (ibid.).
24. Woltosz, e-mail message to author, November 11, 2002.
25. Woltosz, interview by the author, December 22, 2000.
26. Ibid.
27. Ibid. "Wanting more words in his vocabulary. . . . Or things like I need to be able to read a longer lecture . . . . In the past, we had a limited amount of memory. . . . He could store a certain size lecture, and if he had a longer lecture, he had to break it into several pieces. Well, the process of going from one piece to the next could take a couple of minutes with the older computers, let's say ten years ago, so that was a long delay. . . . I've seen him speak to over two thousand people at Berkeley some years ago . . . a two- or three-minute pause in the middle of a talk that has a nice flow to it is a very disruptive kind of event." (Woltosz, interview by the author, December 22, 2000). "In the past, we were limited by the machines, but now, memory is just not an issue anymore. And he hasn't asked for anything like that in a number of years. He's got more than ample space for all of the vocabulary that he needs" (ibid.).
28. Ibid.

29. Woltosz, interview by the author, November 27, 2002.
30. Woltosz, interview by the author, December 22, 2000.
31. Ibid.
32. Woltosz, interview by the author, November 27, 2002.
33. Woltosz, interview by the author, December 22, 2000.
34. Ibid.
35. http://www.damtp.cam.ac.uk/user/Hawking/1996. The British version has disappeared from his Web site: http://www.hawking.org.uk/index.php/disability/disabilityadvice. 03/03/2011. In the same way, in *Black Holes and Baby Universes*, Hawking says, "However, by now I identify with its voice. I would not change even if I were offered a British-sounding voice. I would feel I had become a different person" (26).
36. On the ways in which technologies contribute to the meaning and experiences of the lived body/self and disability, see Debora Lupton and Wendy Seymour, "Technology, Selfhood and Physical Disability," *Social Science and Medicine* 50, no. 12 (2000). According to these authors, people with disabilities avoid the use of certain technologies when they make them look "different." Here the difference (that has become assimilated as part of the self) seems to be both desired and valorized.
37. Woltosz, interview by the author, November 27, 2002.
38. See Nelly Oudshoorn and Trevor Pinch, *How Users Matter: The Co-Construction of Users and Technology* (Cambridge, MA: MIT Press, 2003). According to Intel President Gordon Moore, "This machinery lacked commercial potential and would probably be confined to Hawking's use alone"; see Ron Kampeas, "Hawking Goes Online with New Computer," *Seattle Times*, March 22, 1997. When I asked Woltosz to comment on this, he confirmed, "It is not a mass production system, but a 'one-off' unit constructed just for Hawking" (Woltosz, interview by the author, November 27, 2002).
39. Woltosz, interview by the author, December 22, 2000.
40. Ibid. See also, Moser and Law, "Good Passages, Bad Passages," n. 14.
41. He must surely also use a wireless network on the chair now.
42. Woltosz, interview by the author, December 22, 2000. Woltosz goes on to explain that implementing a new switch is a difficult and time-consuming process that is only undertaken in dire circumstances when Hawking "desperately" needs to change his system; even in such cases, it's hard for Hawking to fit such changes into his schedule (ibid.).
43. Today, both respond to the movement of his eyes.
44. Chris, interview by the author, June 30, 1999.
45. Chris, interview by the author, July 23, 1999. In a certain sense, Hawking becomes their object of research. Put another way, they try to put their feet in his shoes—to become him, trying to feel or guess what he wants.
46. It is interesting to see how Woltosz perceives the assistants' work as purely mechanical, especially when Hawking communicates with his computer. For him, they are just there to put paper in the printer: "Stephen can do so much on his own now . . . in terms of communicating and accessing the world through the Internet" (Woltosz, interview by the author, December 22, 2000).

47. See Harold Garfinkel, *Studies in Ethnomethodology* (Cambridge: Cambridge University Press, 1967), 76–103. On the role of fulfillment in interactions between persons or between persons and machines, see Harry Collins, *Artificial Experts: Social Knowledge and Intelligent Machines* (Cambridge, MA: MIT Press, 1990), and "The Turing Test and Language Skills," in *Technology in Working Order*, ed. Graham Button (London, NY: Sage, 1993), 231–45; Lucy Suchman, *Human-Machine Reconfigurations: Plans and Situated Actions*, 2nd ed. (Cambridge: Cambridge University Press, 2007); Donald MacKenzie, *Knowing Machines: Essays on Technical Change* (Cambridge, MA: MIT Press, 1996), 131–59; Joseph Weizenbaum, "ELIZA: A Computer Program for the Study of Natural Language Communication between Man and Machine," *Communications of the ACM* 9, no. 1 (1966): 36–45; and Sherry Turkle, *Life on the Screen: Identity at the Age of the Internet* (New York: Touchstone, 1997), 102–24.
48. Chris, interview by the author, June 30, 1999.
49. Woltosz, interview by the author, December 22, 2000.
50. See White and Gribbin, *Stephen Hawking*, 121. Gribbin quotes Fischer Dilke, who wrote and directed one of the first television documentaries about Hawking. He says that Hawking would like to be treated in a "normal way"—to "engage in a standard conversation" and not be obliged "to answer only 'Yes' or 'No.'"
51. On the fundamental role of routine as "an intersubjectively shared resource in negotiating daily life," see David Goode, *World without Words* (Philadelphia, PA: Temple University Press, 1994). On the mechanization of the environment, see Ruud Hendriks, "Egg Timers, Human Values and the Care of Autistic Youths," *Science, Technology and Human Values* 23, no. 4 (1998): 399–424.
52. On the concept of performativity, see J. L. Austin, *How to Do Things With Words* (New York: Oxford University Press, 1965).
53. Woltosz, interview by the author, December 22, 2000. Similarly, when I asked his assistant what Hawking's status as head of the General Relativity Group was, he said, "He's heading up, he heads up the group on the super computer, COSMOS Super Computer, and so, if they are making a proposal for funding on how to buy things, he is the one. He doesn't write it, but he'll approve the proposal, and he will sign it. Yeah, 'cause that's what his job is.... *I suppose it's akin to the role of manager director... you're in charge but you don't necessarily do all [the] work*" (Chris, interview by the author, July 23, 1999; my emphasis).
54. Woltosz, interview by the author, November 27, 2002. See also Charles Thorpe and Steven Shapin, "Who Was J. Robert Oppenheimer? Charisma and Complex Organization," *Social Studies of Science* 30, no. 4 (August 2000): 545–90: "Like an ideal early modern gentleman, he was considered to have mastered the art of seeming artless-effortless superiority, *sprezzatura*. He signalled his authority and communicated his expectations, as Wigner describes it, 'very easily and naturally, with just his eyes, his two hands, and a half-lighted pipe'" (574).
55. According to the official job description, the PA is supposed to have "a good general education, preferably to A-level standard, including GCSEs in mathematics and English. Excellent typing skills to 60 wpm, shorthand to 120 wpm (preferable but not essential), and excellent IT skills, including experience in word processing,

e-mail, spreadsheets, and databases. The post-holder should have a mature outlook, be practical, numerate, experienced in interpersonal relationship, and have good negotiating skills. The post-holder will need to demonstrate a high degree of flexibility and be able to use . . . initiative to work totally unsupervised. The post-holder should be able to cope under pressure and have excellent organizational skills [as well as] the ability to drive and travel away from Cambridge (including overseas when necessary). Five years experience at a senior level is required" (Job posting, DAMTP, June 4, 1999).

56. We also see here a symmetry between the way she reads Hawking's face and the first line of his screen; she says that he will respond by yes, no, maybe, or hold. The first line on Hawking's screen is also "clear," "yes," "no," "maybe," "I don't know," "thanks."
57. PA, interview by the author, July 30, 1996.
58. Mostly he will respond with a no, she says, "but it isn't incredibly time-consuming, unless the answer is yes, and . . . that's going to be a lot more time, because . . . every minute detail needs to be gone into" (ibid.).
59. Ibid.
60. I use *preprogrammed* here to illustrate the extreme level of preparation and the constraints that are part of this process. I also use this word to describe the practices of other actors—not just Hawking's.
61. PA, interview by the author, July 30, 1996. One wonders why he can't use instant phrases in this context and to what extent she exaggerates.
62. Ibid.
63. Ibid. (my emphasis).
64. Ibid. (my emphasis).
65. PA, interview by the author, July 1, 1999.
66. David, interview by the author, August 1, 1998.
67. The GA changes every year, though. But, as the secretary mentioned, it's not that simple, because a lot depends on funding at the university.
68. Tom, interview by the author, June 24, 1998.
69. Chris, interview by the author, June 30, 1999.
70. It is interesting to note that the department computer officer in charge of vetting the assistant's computer skills is also in charge of doing the DAMPT's Web page. Hawking's official Web page is linked to the "lay" explanation of black holes for the public. There is no boundary here between esoteric and public understanding.
71. The current GA was at the head of his Department Society, and the previous one was at the head of the Queen May Ball.
72. Chris, interview by the author, June 30, 1999.
73. Tom, interview by the author, June 24, 1998. When I asked the PA if she had an idea of what kind of person should replace her, she said, "I mean, I've got my views, but I don't think that it's particularly relevant because it's Professor Hawking who is going to have the last say on whether it is somebody he thinks he could get on with" (PA, interview by the author, July 1, 1999).
74. Chris, interview by the author, June 30, 1999. As we will see in chapter 2, Hawking

says that he works in the same way—on different problems at the same time (Stephen Hawking, interview by the author, July 1, 1999).

75. Tom, interview by the author, June 24, 1998. As Arthur Koestler says, "When life presents us with a problem it will be attacked in accordance with the code of rules which enabled us to deal with similar problems in the past. These rules of the game range from manipulating sticks to operating with ideas, verbal concepts, visual forms, mathematical entities. When the same task is encountered under relatively unchanging conditions in a monotonous environment, the responses will become stereotyped, flexible skills will degenerate into rigid patterns, and the person will more and more resemble an automaton, governed by fixed habits, whose actions and ideas move in narrow grooves. He may be compared to an engine-driver who must drive his train along fixed rails according to a fixed timetable" (*The Act of Creation* [London: Arkana, 1989], 119).
76. Chris, interview by the author, July 23, 1999.
77. PA, interview by the author, July 1, 1999.
78. Chris, interview by the author, June 30, 1999.
79. Ibid. (my emphasis).
80. Chris, interview by the author, July 20, 1999.
81. Chris, interview by the author, July 23, 1999.
82. Chris, interview by the author, June 30, 1999 (my emphasis).
83. Ibid.
84. Ibid.
85. Ibid. A parallel can be drawn here between the work of translation by the GA and the students, on one hand, and the department computer officer, on the other. See note 70.
86. Ibid. They want Hawking's opinion on everything: "I'm not clear why you need Professor Hawking's support in this matter. He has no experience in this field of home lighting. [. . .] Um, it seems to me that you would be well advised to build a prototype model, and approach a manufacturer with your idea. [. . .] And he has now written back to *me*, very persistently, saying, 'Oh, but I really do want to know what Stephen thinks!' And it's like, you know, I don't care! [laughter] The honest truth is, I don't . . . I'm not going to give you Stephen's opinion on this, because I really don't think that you actually want Stephen's opinion. . . . You want, something else" (ibid.). We witness a similar process with his fan letters.
87. When I asked the GA if he signs with his name or not, he replied, "Well, it's a slight cheat" (Chris, interview by the author, June 30, 1999). Later he will indicate that it is clearly written on the Web. When I checked, it appeared rather more ambiguous to me than this.
88. Ibid.
89. Ibid. Most of the e-mails are sent by his assistants or via his wife. As Woltosz says, "Well, he can use e-mail, but typically my contacts are through Elaine. I think he does more physics-based e-mails, you know, with the people that he's doing his research with, and he lets Elaine or his support staff handle the more background kind of tasks. For him, anything that's not physics is background" (Walt Woltosz,

interview by the author, November 27, 2002). This is what his assistants say as well: they don't see their role.

90. Chris, interview by the author, June 30, 1999.
91. Ibid.
92. Ibid.
93. Ibid. (my emphasis).
94. Ibid. He also mentioned that he had recently put Hawking's e-mail address up on the Web. "It's Stephen's public e-mail address, because people write to me all the time, and say, 'How do I e-mail Stephen?'" Interestingly enough, though they are now writing to Hawking's "official address," it is still the GA who answers.
95. Ibid. "There is a company in Germany . . . who own, (in inverted commas) Stephen's voice, because they actually own the right to that particular sound, because they created the voice, and they own the copyright on the voice. Intel had purchased the voice of X, and had converted it from this card to a piece of software, and that was an example of what it's going to sound like. [. . .] It's been a lot of planning, it's been a lot of legal, . . . but they will send us this voice."
96. Ibid. I use italics when Hawking speaks with the voice synthesizer.
97. Tom, interview by the author, June 24, 1998.
98. Jane Hawking, *Music to Move the Stars: A Life with Stephen* (London: Macmillan, 1999), 523–24 (my emphasis).
99. As Hawking's assistant recalls, "Stephen wanted me to put his chair on charge to do something, but so basically I *had to* be there at that time. [. . .] It was just pretty mundane stuff though. It wasn't anything major, *it would have been major if I hadn't done it*; it was more a preventive measure; it was to stop anything going wrong later, because Stephen was going to the banquet after on his computer battery; obviously, because this is a battery that only lasts a certain amount of time, so we just have to set it up to make sure it works" (Chris, interview by the author, June 30, 1999).
100. Tom, interview by the author, June 24, 1998.
101. CUDO (Cambridge University Development Office) arranged Stephen's White House visit. As Hawking's assistant told me, "It's all about raising the profile of Cambridge University. . . . Although it's got a pretty high profile anyway" (Chris, interview by the author, July 20, 1999).
102. Ibid.
103. Ibid. This statement is from Hawking's inaugural lecture as Lucasian Professor of Mathematics (April 29, 1980).
104. Ibid. Though entirely prepared in advance or already said, the statement must come from 'his mouth.' As the assistant says, he could answer, but his opinion doesn't matter. On the notion of credibility, see also Steven Shapin, *A Social History of Truth: Civility and Science in Seventeenth-Century England* (Chicago: University of Chicago Press, 1994).
105. Chris, interview by the author, July 20, 1999.
106. Ibid.
107. Ibid. When I asked the assistant if he knew why Hawking had agreed to appear on the *Simpsons*, his response was based on a response Hawking had given to a journal-

ist. Chris: "He actually gave an answer to that question in the Sunday *Times*. I can't remember what he said, but I can find out what it was, an exact quote." But later he says, "There were a couple of reasons. It depends who you ask . . . what the answer is" (ibid.).

108. What follows is a transcript of my recording of the press conference.
109. This response seems very similar to Hawking's conference talk entitled "Imagination and Change: Science in the Next Millennium" delivered at the White House: "In 1980 I said I thought there was a 50–50 chance that we would discover a complete unified theory in the next twenty years. We have made some remarkable progress in the period since then, but the final theory seems about the same distance away. [. . .] Nevertheless I am confident we will discover it by the end of the 21st century and probably much sooner. I would take a bet at 50–50 odds that it will be within twenty years starting now." It is thus also similar to his inaugural lecture as Lucasian Professor of Mathematics.
110. This quotation doesn't seem to appear as is anywhere else. I assume it was written from scratch. Hawking had little time, so part of this was probably recycled.
111. Ibid.
112. David Whitehouse, "Hawking Searches for Everything," *BBC News Online*, July 21, 1999, at http://news.bbc.co.uk/2/hi/science/nature/400204.stm. Burt Herman, "Hawking Awaits Unified Theory Proof," *Associated Press Online*, July 21, 1999, online at http://www.highbeam.com/doc/1P1_23166588.html; Stephen E. Jones, "Re: Hawking Awaits Unified Theory Proof," August 6, 1999, online at http://www.asa.chm.colostate.edu/archive/evolution/19990810035.html; "Hawking Awaits Unified Theory Proof," *Florida Today*, July 23, 1999, online at http://www.floridatoday.com/space/explore/stories/1999b/072399b.htm.
113. See Bruno Latour and Steve Woolgar, *Laboratory Life: The Social Construction of Scientific Facts* (Beverly Hills, CA: Sage, 1979).
114. One can draw a parallel between the program utilized by Hawking, such as EZ Keys, which can repeat phrases as many times as one wants (which is to say, as long as the interlocutor has integrated what is said), and the media machine, which repeats the same statement indefinitely: "Very often, when we say something verbally, our communication partner, who wasn't really paying attention, says, 'Huh?' So you have to say it again, and so long as you haven't started a new sentence, you can press it again. Suppose I wanted to say, 'What are you going to do tomorrow morning?' I repeat the process" (Woltosz, interview by the author, December 22, 2000).
115. *Deseret News* (Salt Lake City), July 21, 1999.
116. See Erving Goffman regarding his notion of the territoriality of the self or how things become an extension of the self (*The Presentation of Self in Everyday Life* [Garden City, NY: Doubleday, 1959]).
117. Woltosz had to take the place of its future user when he invented the program (in much the same way that Hawking uses diagrams to understand how black holes function, as we will see in chapter 3). "If you can know what your user wants to do, and you can sit down and practice doing that, then you will find the things that are

aggravating or inefficient . . . just not very well done. You will find ways to improve it" (Woltosz, interview by the author, December 22, 2000). On the inscription of the representation of the user in the artifact, see Madeleine Akrich, "The De-scription of Technical Objects," in *Shaping Technology/Building Society: Studies in Sociotechnical Change*, ed. Wiebe Bijker and John Law (Cambridge, MA: MIT Press, 1992), 205–24.

118. The vocabulary is adapted to someone who has the same disability (e.g., "Can I have suction please!"). The computer can also memorize a vocabulary specific to Hawking (e.g., he can add to or remove from his list of words the names of his assistants, the places he goes, the people he meets).

119. Without this collective of competencies, there is no HAWKING. At the same time, there is something very specific to Hawking's personality that makes this collective function, and of course these immutable mobiles are not totally immutable, insofar as slight modifications occur that will be standardized again.

120 In Hawking's case, it is interesting to see the delegation of competencies to humans and machines, but also the delegation of roles. In other words, the cultivated nurses will have the conversation for him; his first wife mentioned that she felt like she was the mother *and* the father of her children. Everything is redistributed in his environment: the roles, the cognitive capacities, the corporeal tasks. But it is through this extreme distribution that a singularity emerges. In comparison with other disabled persons, he has a level of care that others do not; he is treated differently; he has to work more than an average (i.e., not-so-famous) disabled man.

121. Russell Shuttleworth: "I must emphasize again, however, that an intersubjective dynamic yet exists within the simple routine that makes it qualitatively different from the incorporated bodily extension of the tool. This intersubjectivity, however, can sometimes break down when the personal assistant lapses back into the habits of his own embodied intentionality. For instance, several times late at night when I am tired and am about to brush the teeth of one of my employers, upon grasping the toothbrush and putting some toothpaste on it, I have stuck it in my own mouth by mistake" ("The Pursuit of Sexual Intimacy for Men with Cerebral Palsy," PhD diss., University of California, San Francisco and Berkeley, 2000, 106).

122. This form of resistance is also typical of others in the same situation; see, for example, Albert Robillard, *Meaning of a Disability: The Lived Experience of Paralysis* (Philadelphia, PA: Temple University Press, 1999); and Michel Callon and Vololona Rabeharisoa, "Gino's Lesson on Humanity: Genetics, Mutual Entanglements and the Sociologist's Role," *Economy and Society* 33, no. 1 (2004): 1–27.

123. The form of resistance he manifests could be analyzed in terms similar to what Laurent Thévenot describes as a "regime of familiarity"; see "Pragmatic Regimes Governing the Engagement with the World," in *The Practice Turn in Contemporary Theory*, ed. Theodore R. Schatzki, Karin Knorr Cetina, and Eike von Savigny (London: Routledge, 2001), 56–74.

124. PA, interview by the author, July 30, 1996.

125. Gribbin and White, *Stephen Hawking*, 268. I describe more specifically in chapter 5 how Hawking's way of interacting can be interpreted as a lack of politeness

(e.g., how he manifests his anger by rolling over the feet of those who anger—or frustrate—him).

126. Chris, interview by the author, June 30, 1999.

Chapter Two

1. Kip S. Thorne, *Black Holes and Time Warps: Einstein's Outrageous Legacy* (New York: W. W. Norton, 1994), 419 (my emphasis).
2. On the role and importance of the manipulation of inscriptions in scientific work, see, for example, Bruno Latour, "Visualization and Cognition: Thinking with Eyes and Hands," in *Knowledge and Society: Studies in the Sociology of Culture Past and Present* 6 (1986): 1–40; David Kaiser, *Drawing Theories Apart: The Dispersion of Feynman Diagrams in Postwar Physics* (Chicago: University of Chicago Press, 2005); Andrew Warwick, *Masters of Theory: Cambridge and the Rise of Mathematical Physics* (Chicago: University of Chicago Press, 2003).
3. Brian Rotman, *Mathematics as Sign: Writing, Imagining, Counting* (Stanford, CA: Stanford University Press, 2000), 34.
4. Andrew Warwick, for example, has examined the mathematical "wranglers" and the study of mathematics at Cambridge University in the eighteenth-century before paper and pen were used; see his *Masters of Theory*.
5. Stephen Hawking, online at http://www.hawking.org.uk/index.php/about-stephen/questionsandanswers (my emphasis).
6. John Gribbin and Michael White, *Stephen Hawking: A Life in Science* (New York: Dutton, 1992), 60 (my emphasis).
7. On the false dichotomy between theory and practice, see Bruno Latour, "Sur la pratique des théoriciens," in *Savoir théoriques et savoirs d'action*, ed. Jean-Marie Barbier (Paris: Presses Universitaires de France, 1998), 131–46. David Kaiser calls this "the practice of theory"; see his *Drawing Theories Apart*. See also Warwick, *Masters of Theory*, particularly chapter 1.
8. Regarding this point, see Hélène Mialet, "Do Angels Have Bodies? Two Stories about Subjectivity in Science: The Cases of William X and Mister H," *Social Studies of Science* 29, no. 4 (1999): 551–82.
9. Here one could perhaps cite the entire corpus of recent work in science and technology studies—there is near unanimous agreement in the discipline about this. However, the work that has been done regarding the collectivization and materialization of scientific knowledge is far more developed with regard to experimental science than theoretical science. On this point, see Warwick, *Masters of Theory*, esp. chapter 1; and Andrew Pickering and Adam Stephanides, "Constructing Quaternions: On the Analysis of Conceptual Practice," in *Science as Practice and Culture*, ed. Andrew Pickering (Chicago: University of Chicago Press, 1992).
10. Lyndell Bell, interview by the author, June 12, 1996.
11. Neil Turok, interview by the author, July 1, 1999.
12. Raymond, interview by the author, March 7, 1999. This student also wonders how

Hawking can mobilize information. He says that he can page through newspapers in airports, for example, whereas Hawking cannot do that.

13. Concerning the metaphor of the body, see the description of Pasteur in Michel Callon and John Law, "After the Individual in Society: Lessons on Collectivity from Science, Technology and Society," *Canadian Journal of Sociology* 22, no. 2 (1950): 169. The published version is worded differently; I prefer the draft version. See also Michel Callon and John Law, "Agency and the Hybrid Collective," *South Atlantic Quarterly* 94, no. 2 (1995): 481–507; and Bruno Latour, *The Pasteurisation of France* (Cambridge, MA: Harvard University Press, 1988).
14. For instance, Albert Robillard, who is a professor and also has ALS, has exactly the same schedule. He arrives at his office at 11:00 a.m. because of his morning treatment.
15. Reginald Golledge, "On Reassembling One's Life: Overcoming Disability in the Academic Environment," *Environment and Planning D: Society and Space* 15, no. 4 (1997): 391–409. See also, Robert Murphy, *The Body Silent* (New York: W. W. Norton, 1987).
16. Bernard Carr, interview by the author, July 16, 1996.
17. Gribbin and White, *Stephen Hawking*, 158. One can see how the boundary between personal and professional is blurred. See also Steven Shapin, *A Social History of Truth: Civility and Science in Seventeenth-Century England* (Chicago: University of Chicago Press, 1994), chap. 8.
18. Raymond, interview by the author.
19. PA, interview by the author, July 30, 1996 (my emphasis).
20. Raymond, interview by the author. In chapter 6 we'll see that letters from Princeton have been found in the archives. Brandon Carter has confirmed the important role of John Wheeler at Princeton (interview by the author, April 11, 2007).
21. Raymond, interview by the author.
22. At the time of my interview, the archive was based at Los Alamos; now it is at Cornell University.
23. Turok, interview by the author (my emphasis).
24. Raphael, interview by the author, July 17, 1996.
25. It seems, however, that there is still delegation of competencies from person to person: "So, I think today, he'd probably . . . do it by himself, using the World Wide Web, instead of having somebody showing them to him; although he might use somebody, because it might be faster" (Raymond, interview by the author).
26. Thomas, interview by the author, June 30, 1998. And as she mentions, "I haven't a fixed idea of what I think of him now, his public figure" (Thomas, interview by the author). However, they are all aware that Hawking's fame will help them in the job market, especially if they leave physics.
27. Tim, interview by the author, June 24, 1998.
28. Ibid.
29. Similar practices have been well documented with regard to experimental physics. As Kathryn Olesko has shown, for example, in the case of the German physicist Friedrich Kohlrausch, "[The students'] roles in the investigation were as partners

in research; to a certain degree there was a temporary fusion of their identities as researchers, as Kohlrausch's reports indicate (while the "voice" of reporting was Kohlrausch's, the performance was completed in tandem). The student thus had the opportunity to act as a researcher before he actually was one; as in teaching, acting "as if" in research was an important step toward *becoming* a researcher." See Kathryn Olesko, "The Foundations of a Canon: Kohlrausch's Practical Physics," in *Pedagogy and Practice of Science: Historical and Contemporary Perspectives*, ed. D. Kaiser (Cambridge, MA: MIT Press, 2005), 323–57, at 341.

30. On the crucial role of collaboration and personal communication among theoretical physicists, see David Kaiser, "Making Tools Travel: Pedagogy and the Transfer of Skills in Postwar Theoretical Physics," in *Pedagogy and Practice of Science: Historical and Contemporary Perspectives*, ed. D. Kaiser (Cambridge, MA: MIT Press, 2005), 325.
31. Turok, interview by the author.
32. Hubert Dreyfus, *What Computers Can't Do: A Critique of Artificial Reason* (New York: Harper & Row, 1972), and "Why Computers Must Have Bodies in Order to Be Intelligent," *Review of Metaphysics* 21 (September 1967): 13–32.
33. In this sense, he is the opposite of the solitary researcher. As one of his former students comments, "I think quite often scientists are . . . too much in 'ivory towers.' [. . .] Because of his handicap, and this . . . fact that the students are helping, it gives them a special relationship, and then you can . . . take advantage of this, in a way that it would be much harder with some other people" (Raymond, interview by the author).
34. Tim, interview by the author.
35. On the indispensable (and often, though not always, invisible) role of technicians in the process of knowledge production see Albert Robillard, *Meaning of a Disability: The Lived Experience of Paralysis* (Philadelphia, PA: Temple University Press, 1999); Shapin, *Social History of Truth*, chap. 8; Otto Sibum, "Reworking the Mechanical Equivalent of Heat," *Studies in History and Philosophy of Science* 26, no. 1 (1995): 73–106; Michael Lynch, Eric Livingston, and Harold Garfinkel, "Temporal Order in Laboratory Work," in *Science Observed*, ed. Karin Knorr Cetina and M. Mulkay (London: Sage, 1983), 205–38; Karin Knorr Cetina, *Epistemic Cultures: How the Sciences Make Knowledge* (Cambridge, MA: Harvard University Press, 1999).
36. Regarding the history of mathematical practices peculiar to Cambridge University, see Warwick, *Masters of Theory*.
37. Tim, interview by the author.
38. Christophe, interview by the author, October 11, 2006. In this particular context, one can make a link with what Turok says about Hawking: "Yes, what you need is a very clear understanding of the principles, and that's what he has. And so those same principles can be used again and again and again in different contexts" (Turok, interview by the author).
39. Raphael, interview by the author.
40. Tim, interview by the author.
41. On the role of analogical reasoning in Hawking's cosmological thinking, see Gu-

staaf C. Cornelis, "Analogical Reasoning in Modern Cosmological Thinking," in *Metaphor and Analogy in the Sciences*, ed. F. Hallyn (Netherlands: Kluwer Academic Publishers, 2000), 165–80.

42. Thomas Kuhn, *The Structure of Scientific Revolutions* (Chicago: University of Chicago Press, 1962); Michael Polanyi, *The Tacit Dimension* (Garden City, NY: Doubleday, 1966).
43. For an examination of the links between different types of training and the resulting scientific practices, see David Kaiser, ed., *Pedagogy and Practice of Science: Historical and Contemporary Perspectives* (Cambridge, MA: MIT Press, 2005).
44. Christophe, interview by the author, July 2, 2005 (my emphasis).
45. Black hole percolation, quantum cosmology, information loss, and string theory are all problems of quantum gravitation; that is, they are trying to unite general relativity with quantum mechanics.
46. See note 38 above. Moreover, he has been focusing on the same subjects for a very long time. See Gary Gibbons and Stephen Hawking, *Euclidean Quantum Gravity* (Singapore: World Scientific Publishing, 1993).
47. On the creation, the role, and the persistence of a canonical text in science pedagogy, see the important article by Olesko, "Foundations of a Canon," 323–57, esp. 325.
48. Tim, interview by the author (my emphasis).
49. Thomas's research is an extension of Raphael's work on quantum cosmology.
50. Thomas, interview by the author.
51. Ibid. (my emphasis).
52. Ibid.
53. I refer here to the famous distinction between the context of discovery and the context of justification borrowed from Reichenbach and applied by Karl Popper in *The Logic of Scientific Discovery* (London: Hutchinson, 1972) and *Conjectures and Refutations: The Growth of Scientific Knowledge* (London: Routledge and Kegan Paul, 1969).
54. See for example, Jack Goody, *The Domestication of the Savage Mind* (Cambridge: Cambridge University Press, 1977).
55. Raphael, interview with the author (my emphasis).
56. Hawking says that he uses this program on his Web page, but this doesn't seem to be the case. His secretary uses it when she works for other professors in the department, which again allows us to see Hawking as kind of window into the practices of other theoreticians.
57. On the organization of research schools, see "Research Schools, Historical Reappraisals," ed. Gerald L. Geison and Frederic L. Holmes, special issue of *Osiris*, 2nd series, 8 (1993). We see here the hierarchical relation between the professor "who has ideas" and the technicians who do the work of detailed calculation and proof. This echoes Popper's distinction between the context of discovery—the realm of free imagination, where ideas come from, and the context of justification (see n. 53 above), where they are tested (falsified). Similarly, Lorraine Daston has shown how, in the nineteenth century, imagination and calculation were refracted through

notions of gender. Talent and genius, she argues, became associated with imagination; calculation, on the other hand, was demoted to the automatic and the rote. "Calculation took on the dull, patient associations of repetitive and ill-paid bodily labor, ranked as the lowest of the mental faculties. Hence it comes as no surprise that women, once scorned for their vivid imaginations and mental restlessness, ultimately staffed the *bureaux de calculs* that did the plodding work of compiling the tables and reducing the data for major astronomical and statistical projects until the end of World War II." "Enlightenment Calculation," *Critical Inquiry* 21, no. 1 (1994): 182–202, at 186. See also Lorraine Daston, "Condorcet and the Meaning of Enlightenment," *Proceedings of the British Academy* 151 (December 2007): 113–34. While this is clearly not the case for Hawking's students, who are considered exceptional, it sets the established hierarchy between students and professors, bodies and minds, in sharp relief.

58. Raymond, interview by the author.
59. Raphael, interview by the author.
60. Penrose, interview by the author, June 17, 1998.
61. Thomas, interview by the author.
62. Tim, interview by the author.
63. Thomas, interview by the author.
64. We will see how this work operates differently in the chapters of this book.
65. Moreover, because of the conditions his disability imposes, he has to visualize calculations rather than doing them (see chapter 3). This is also the only way "we," "the public," can comprehend what he does, as Gribbin and White mention: "His work is a major part of Stephen Hawking, but so few of us can understand it except in the vaguest pictorial terms" (*Stephen Hawking*, 292).
66. Thomas, interview by the author.
67. Tim, interview by the author. Penrose, by contrast, finds it difficult to work with Hawking because of his way of communicating. The slowness of the process when he writes and his intervention "break" Penrose's "train of thought" (Penrose, interview by the author, June 17, 1998).
68. Thomas, interview by the author (my emphasis).
69. Ibid.
70. Ibid.
71. Kip Thorne, interview by the author, August 26, 1999 (my emphasis).
72. Turok, interview by the author.
73. Thomas, interview by the author.
74. Christophe, interview by the author, July 2, 2005.
75. Ibid.
76. Stephen Hawking, interview by the author, July 7, 1999. On the role of metaphor (e.g., "getting stuck"), see Georges Lakoff, *Women, Fire, and Dangerous Things: What Categories Reveal About the Mind* (Chicago: University of Chicago Press, 1987).
77. Christophe, interview by the author, October 11, 2006: "I didn't have a specific subject in mind, but I knew I wanted to work on something quite philosophical and quite deep, and he suggested something even more difficult than what I ended up

working on. But I'm sure he did the same thing with the last ten students he worked with. They probably all accepted. But in fact we all end up working on something easier, it's too huge, too general."

78. Raymond, interview by the author (my emphasis).
79. Raphael, interview by the author.
80. Ibid. (my emphasis).
81. Christophe, interview by the author, October 11, 2006.
82. See note 13.
83. Woltosz, interview by the author, November 27, 2002.

Chapter Three

1. Jim Hartle, interview by the author, June 1999.
2. Ibid.
3. Kip Thorne, interview by the author, August 26, 1999.
4. Ibid.
5. David Filkin, *Stephen Hawking's Universe: The Cosmos Explained*, with a foreword by Stephen Hawking (New York: Basic Books, 1998), xiv (my emphasis).
6. According to Louis Quéré, the neo-Cartesian paradigm for cognition could be summarized as follows: cognition is lodged in the mind, it is the product of the manipulation of internal representations of the external word, and to think is to calculate. Hawking seems to confirm the two first principles, but distances himself from the third. On the neo-Cartesian paradigm and its critics, see Louis Quéré, "La situation toujours négligée," *Réseaux* 15, no. 85 (1997): 163–92. On the history of the complex relationship between calculation and mental faculty, see Lorraine Daston, "Enlightenment Calculation," *Critical Inquiry* 21, no. 1 (1994): 182–202.
7. H. M. Collins, "The Seven Sexes: A Study in the Sociology of a Phenomenon, or the Replication of Experiment in Physics," *Sociology* 9, no. 2 (1975): 205–24; and *Changing Order: Replication and Induction in Scientific Practice*, 2nd ed. (Chicago: University of Chicago Press, 1992); Peter Galison, *How Experiments End* (Chicago: University of Chicago Press, 1987); Simon Schaffer, "Glassworks: Newton's Prism and the Uses of Experiment," in *The Uses of Experiment: Studies in the Natural Sciences*, ed. David Gooding, Trevor Pinch, and Simon Schaffer (Cambridge: Cambridge University Press, 1989), 67–104.
8. Regarding this point, see Brian Rotman's critique of the traditional characterization of mathematics as a form of Platonic realism (i.e., mathematical objects are mentally accessible, but nevertheless owe nothing to human culture), and Lej Brouwer's conception of mathematics as "a languageless activity." See Brian Rotman, "Thinking Dia-grams: Mathematics, Writing, and Virtual Reality," *South Atlantic Quarterly* 94, no. 2 (1995): 391.
9. On this point, see Ursula Klein, ed., *Tools and Modes of Representation in the Laboratory Sciences* (Dordrecht, Holland: Kluwer Academic Publishers, 2001); David Kaiser, *Drawing Theories Apart: The Dispersion of Feynman Diagrams in Postwar Physics* (Chicago: University of Chicago Press, 2005); Andrew Warwick, *Masters of Theory:*

Cambridge and the Rise of Mathematical Physics (Chicago: University of Chicago Press, 2003); Andrew Pickering, "Against Putting the Phenomena First: The Discovery of the Weak Neutral Current," *Studies in History and Philosophy of Science* 15, no. 2 (1984): 85–117; and *The Mangle of Practice: Time, Agency, and Science* (Chicago: University of Chicago Press, 1995); Andrew Pickering and Adam Stephanides, "Constructing Quaternions: On the Analysis of Conceptual Practice," in *Science as Practice and Culture*, ed. Andrew Pickering (Chicago: University of Chicago Press, 1992); Elinor Ochs, Sally Jacoby, and Patrick Gonzales, "Interpretive Journeys: How Physicists Talk and Travel Through Graphic Space," *Configurations* 2, no. 1 (1994): 151–71; Reviel Netz, *The Shaping of Deduction in Greek Mathematics* (Cambridge: Cambridge University Press, 1999); Claude Rosental, *Weaving Self-Evidence: A Sociology of Logic*, trans. Catherine Porter (Princeton, NJ: Princeton University Press, 2008); Christian Jacob, ed., *Lieux de savoir: Les mains de l'intellect* (Paris: Albin Michel, 2011).

10. Christophe, interview by the author, July 1, 2005.
11. Brandon Carter, interview by the author, April 11, 2007 (my emphasis).
12. Stephen Hawking, online at http://www.hawking.org.uk/index.php/about_stephen/questionsandanswers.
13. The words are given to him by the machines or by the students.
14. According to Kip Thorne, "Penrose, Hawking, Robert Geroch, Georges Ellis, and other physicists created a powerful set of combined topological and geometrical tools for general relativity calculations, tools that are now called *global methods*"; see his *Black Holes and Time Warps : Einstein's Outrageous Legacy* (New York: W. W. Norton, 1994), 465.
15. Christophe, interview by the author, July 8, 2005.
16. See, for example, Bruno Latour and Steve Woolgar, *Laboratory Life: The Social Construction of Scientific Facts* (Beverly Hills: Sage, 1979); Michael Lynch, "Discipline and the Material Form of Images: An Analysis of Scientific Visibility," *Social Studies of Science* 15, no. 1 (1985): 37–66; and "The Externalized Retina: Selection and Mathematization in the Visual Documentation of Objects in the Life Sciences," *Human Studies* 11, nos. 2–3 (1988): 201–34; Michael Lynch and Steve Woolgar, eds., *Representation in Scientific Practice* (Cambridge, MA: MIT Press, 1990).
17. Michael Wintroub makes a similar point with regard to the ability of royal rituals and cabinets of curiosity to make the universe metonymically available to the king's gaze; see his *A Savage Mirror: Power, Identity and Knowledge in Early Modern France* (Stanford, CA: Stanford University Press, 2006), 172.
18. On the concept of infinity, see Brian Rotman, *Ad Infinitum . . . the Ghost in Turing's Machine: Taking God out of Mathematics and Putting the Body Back In* (Stanford, CA: Stanford University Press, 1993).
19. Alan Guth, interview by the author, March 10, 2006. Brandon Carter recalls that these diagrams borrow principles developed by Mercator. Brandon Carter, interview by the author, April 11, 2007.
20. Christophe, interview by the author, February 16, 2007.
21. Alan Guth, interview by the author, March 10, 2006.
22. Christophe, interview by the author, July 8, 2005.

23. Christophe, interview by the author, February 16, 2007. The diagram seems to be used as a heuristic tool (i.e., as a way of asking the right questions) as much as it is used in the resolution of problems.
24. On the role of artifacts in cognitive performances, see Bruno Latour, "Le travail de l'image ou l'intelligence redistribuée," *Culture Technique*, no. 22 (1991): 12–24; Lucy A. Suchman, *Human-Machine Reconfigurations: Plans and Situated Actions*, 2nd ed. (Cambridge: Cambridge University Press, 2007); Edwin Hutchins, "How a Cockpit Remembers Its Speeds," *Cognitive Science* 19, no. 3 (1985): 265–88; and *Cognition in the Wild* (Cambridge, MA: MIT Press, 1995); Charles Goodwin, "Practices of Seeing, Visual Analysis: An Ethnomethodological Approach," in *Handbook of Visual Analysis*, ed. Theo Van Leeuwen and Carey Jewitt (London: Sage, 2000), 157–82; Donald Norman, *Things That Make Us Smart: Defending Human Attributes in the Age of the Machine* (Reading, MA: Addison-Wesley, 1993).
25. Gilles Châtelet, *Figuring Space: Philosophy, Mathematics and Physics* (Dordrecht, Holland: Kluwer Academic Publishers, 2000); see especially Kenneth J. Knoespel's introduction.
26. Ibid., 10.
27. Christophe, interview by the author, July 8, 2005 (my emphasis).
28. David Kaiser, "Stick-Figure Realism: Conventions, Reification, and the Persistence of Feynman Diagrams, 1948–1964," *Representations* 70 (Spring 2000): 62. On the role and history of pedagogy in physics, see Kaiser, *Drawing Theories Apart*, and Warwick, *Masters of Theory*.
29. "Most people who don't know how to read a diagram can learn very quickly. It's not 'sorcery'; once you have a diagram before your eyes, it's not as complicated as we imagine it; it's quite simple" (Christophe, interview by the author, July 8, 2005).
30. Christophe, interview by the author, February 16, 2007.
31. Christophe, interview by the author, October 11, 2006 (my emphasis).
32. Christophe, interview by the author, July 8, 2005.
33. On a similar form of inversion, or how the public geometrical form of knowledge becomes private, see Peter Galison, "The Suppressed Drawing: Paul Dirac's Hidden Geometry," *Representations* 72 (Autumn 2000): 145–66.
34. Though Hawking says that he doesn't use Feynman diagrams because they are not useful for his research, one wonders if perhaps he doesn't do research in this domain because he can't use them easily.
35. See Kaiser, *Drawing Theories Apart*.
36. Leibniz struggled with similar issues of how to find a written means of making immediately—and visually—present something otherwise obscured by long strings of calculation. See Matthew Jones, *The Good Life in the Scientific Revolution* (Chicago: University of Chicago Press, 2006). I would like to thank David Kaiser for pointing this out to me.
37. Brian Rotman, "Figuring Figures" (paper presented at an MLA round table entitled "Between Semiotics and Geometry: Metaphor, Science, and the Trading Zone," Philadelphia, PA, December 28, 2004), 10 (my emphasis).
38. Kaiser, interview by the author, March 9, 2006.

39. As Sam Schweber pointed out to me, "Feynman diagrams considered as tools are local methods, in the sense of Kip Thorne. They are visual translations of perturbative methods. Their usefulness rests primarily on the ability of the theorist using them to calculate their (numerical) contribution to the process under consideration. Though one can readily draw the diagrams that contribute to any order of perturbation theory—only summing their contributions to a given order gives you insight into the process. Thus Kinoshita's amazing prowess is his ability to make powerful computers do the evaluation of the thousand or so Feynman diagrams he had to consider to calculate the electron's magnetic moment to 8th order in perturbation theory. Clearly Hawking could not do that" (personal e-mail communication to the author, July 8, 2009).
40. Christophe, interview by the author, July 8, 2005.
41. Christophe, interview by the author, October 11, 2006.
42. Alan Guth, interview by the author, March 2005.
43. Felicity Mellor, interview by the author, March 2001.
44. Rotman, "Figuring Figures," 3.
45. Christophe, interview by the author, October 11, 2006 (my emphasis).
46. Ibid.
47. Ochs, Jacoby, and Gonzales, "Interpretive Journeys." On virtual experimentation or, more specifically, the idea of virtual witnessing, see Steven Shapin and Simon Schaffer, *Leviathan and the Air-Pump: Hobbes, Boyle, and the Experimental Life* (Princeton, NJ: Princeton University Press, 1989); Michael Wintroub, "The Looking Glass of Facts: Collecting, Rhetoric and Citing the Self in the Experimental Natural Philosophy of Robert Boyle," *History of Science* 35 (June 1997): 189–217.
48. Christophe, interview by the author, July 8, 2005.
49. Châtelet, *Figuring Space*, 11.
50. Woltosz, interview by the author, December 22, 2000.
51. Christophe, interview by the author, July 8, 2005.
52. This contradicts what Kuhn says about thought experiments—that is, he argues that this doesn't require the work of the body. On the indispensable role of the body and instruments in thought experiments and on the thought experiment as experimental device, see David C. Gooding, "What Is Experimental about Thought Experiments?" in *Proceedings of the Biennial Meetings of the Philosophy of Science Association* 2, no. 2 (1992): 280–90. As Brian Rotman points out, "Diagrams—whether actual figures drawn on the page or their imagined versions—are *the work of the body*; they are created and maintained as entities and attain significance only in relation to *human visual-kinetic presence*, only in relation to our experience of the cultural inflicted world. As such, they not only introduce the historical contingency inherent to all cultural activity, but, more to the present point, *they call attention to the materiality of all signs and of [the] corporeality of those who manipulate them*. [...] In other words, *diagrams are inseparable from perception: only on the basis of our encounters with actual figures can we have any cognitive or mathematical relation to their idealized forms*" (my emphasis). See Rotman, "Thinking Dia-grams," 401, and "Figuring Figures"; see also Châtelet, *Figuring Space*.

53. Ochs, Jacoby, and Gonzales, "Interpretive Journeys," 152, 164.
54. Ibid., 170–71. On the process of identification of the researcher with his object of research, see also Hélène Mialet, *L'Entreprise créatrice: Le rôle des récits, des objets et de l'acteur dans l'invention* (Paris: Hermès-Lavoisier, 2008); and Evelyn Fox Keller, *A Feeling for the Organism* (San Francisco: W. H. Freeman, 1983).
55. Tim, interview by the author, June 24, 1998.
56. On the relationships between construction and realism, see Kaiser, "Stick-Figure Realism." See also Lynch, "Discipline and the Material Form of Images." On models, see Ian Hacking, *Representing and Intervening: Introductory Topics in the Philosophy of Natural Science* (New York: Cambridge University Press, 1983).
57. Christophe, interview by the author, February 16, 2007 (my emphasis).
58. On this point, see Harry M. Collins, *Artificial Experts: Social Knowledge and Intelligent Machines* (Cambridge, MA: MIT Press, 1990), chap. 5.
59. There is here an interesting comparison to make with Feynman's keen sense of visualization as he described it to Sam Schweber: "One delightful example that I really got big pleasure out of, is the liquid helium problem that I like because I spent a long time picturing that damn thing and doing everything without writing any equations, and it was one time in my life where I did an awful lot of physics *with my hand tied behind my back* because I didn't know how to write a damn thing, and there was nothing I could do but keep on picturing, . . . and I couldn't get anything down mathematically" (my emphasis). See Silvan S. Schweber, *QED and the Men Who Made It: Dyson, Feynman, Schwinger, and Tomonaga* (Princeton, NJ: Princeton University Press, 1994), 466. See also Roger Penrose, *The Emperor's New Mind: Concerning Computers, Mind and the Laws of Physics* (New York: Penguin Books, 1989), especially 548–50.
60. Charles Sanders Peirce, quoted in Brian Rotman, "The Technology of Mathematical Persuasion," in *Inscribing Science: Scientific Texts and Materiality of Communication*, ed. Timothy Lenoir (Stanford, CA: Stanford University Press, 1998), 65. See also Nancy Nersessian, "In the Theoretician's Laboratory: Thought Experimenting as Mental Modeling," *Proceedings of the Biennial Meeting of the Philosophy of Science Association* 2 (1992): 291–301. "Although no instructions are given to imagine or picture the situations, when queried about how they had made inferences in response to the experimenter's questioning, most subjects report it was by means of 'seeing' or 'being' in the situation depicted. That is, the reader sees herself as an 'observer.' Whether the view of the situation is 'spatial,' i.e., a global perspective, or 'perspectival,' i.e., from a specific point of view, is currently under investigation," 295.
61. Rotman, "Thinking Dia-grams," 397.
62. It is not because the result is exact that he stops sending them away to do the calculations, but because he stops sending them away to do the calculation that the result becomes true.
63. Modernity is often thought of in terms of spatial and temporal distance. On the one hand, we today are supposed to be much more advanced than we once were; that is, our cognitive competencies are deemed much more sophisticated than they were in the past. On the other hand, Western societies are adduced to be much more ad-

vanced than "primitives" in the wild. Because Hawking cannot write and draw with pen and paper, his practice is not dissimilar from that of either Aquinas, who composed his ideas—the *Summa Theologica*—in his head through elaborate techniques of memory, or from an Amazonian tribesman today, who relies upon his memory to find his path. See, for example, Mary Carruthers, *The Book of Memory: A Study of Memory in Medieval Culture* (Cambridge: Cambridge University Press, 2008); and Jack Goody, *The Domestication of the Savage Mind* (Cambridge: Cambridge University Press, 1977).

64. Christophe, interview by author, February 16, 2007.

Chapter Four

1. Roland Barthes, *Mythologies* (New York: Hill and Wang, 1987), 68–69.
2. This reservoir of texts and tropes functions in and across popular discourse by means of media that work differently as forms for storage and transmission: the hard drive, print publication, television, cinema." I would like to thank Lisa Gitelman for this comment.
3. Certain details concerning his private life and the role of his nurses remain. The computer is generally perceived more as a means of communication than as a tool for working.
4. On this point, see Schaffer, "Genius in Romantic Natural Philosophy," in *Romanticism and the Sciences*, ed. Andrew Cunningham and Nick Jardine (Cambridge: Cambridge University Press, 1990), 82–98 ; and William Clark, "On the Ironic Specimen of the Doctor Philosophy," *Science in Context* 5, no. 1 (1992): 97–137.
5. Jeremy Dunning-Davies, "Popular Status and Scientific Influence: Another Angle on 'The Hawking Phenomenon,'" *Public Understanding of Science* 2, no. 1 (1993): 85–86.
6. Arthur Lubow, "Heart and Mind," *Vanity Fair*, June 1992, 44–53, paraphrasing from 53.
7. Ian Ridpath, "Black Hole Explorer," *New Scientist*, May 4, 1978, 307.
8. For an analysis of the repertoires "for speaking about the bodily circumstances that either assisted or handicapped the processes by which genuine knowledge was to be attained," see *Science Incarnate: Historical Embodiments of Natural Knowledge*, ed. Lawrence and Shapin (Chicago: University of Chicago Press, 1998), especially Steven Shapin's contribution, "The Philosopher and the Chicken: On the Dietetics of Disembodied Knowledge," 21–51, and Janet Brown's, "I Could Have Retched All Night: Charles Darwin and His Body," 240–88.
9. Ridpath, "Black Hole Explorer," 308, n. 7. But the fact that it is science remains essential. As another scientific journalist told me, "I don't expect that you would be doing the same project if Hawking had been a composer writing music. It would be interesting if Hawking had been writing music in his head and dictating to somebody or a computer. That would not be the same thing; people think science is different from other aspects of culture" (John Gribbin, interview by the author, September 16, 1997).

10. Bryan Appleyard, "A Master of the Universe," *Sunday Times Magazine* (London), June 19, 1988, 26.
11. Jerry Adler, Gerald C. Lubenow, and Maggie Malone, "Reading God's Mind," *Newsweek*, June 13, 1988, 36 (my emphasis).
12. The fact of seeing the inventor through his or her invention is, I argue, a process of attribution constitutive of invention; on this point see Hélène Mialet, *L'Entreprise créatrice: Le rôle des récits, des objets et de l'acteur dans l'invention* (Paris: Hermès-Lavoisier, 2008), and "Making a Difference by Becoming the Same," *International Journal of Entrepreneurship and Innovation* 10, no. 4 (2009): 257–65.
13. Regarding the collective construction of the genius, see Geoffrey Cantor, "The Scientist as Hero: Public Images of Michael Faraday," in *Telling Lives in Science: Essays on Scientific Biography*, ed. Richard Yeo and Michael Shortland (Cambridge: Cambridge University Press, 1996), 171–95; Richard Yeo, "Genius, Method, and Morality: Images of Newton in Britain, 1760–1860," *Science in Context* 2, no. 2 (1988), 257–84; Nathalie Heinich, *La gloire de Van Gogh: Essai d'anthropologie de l'admiration* (Paris: Editions de Minuit, 1991).
14. See Stephen Hawking, ed., *Qui êtes vous Mr Hawking?* (Paris: Odile Jacob, 1994), 204. This book was billed as *A Reader's Companion* to *A Brief History of Time*, the movie directed by Errol Morris; it was edited by Hawking and prepared by Gene Stone. I take my quotations from the English-language edition, published in the United States as Stephen Hawking, ed., *A Brief History of Time: A Reader's Companion*, prepared by Gene Stone (New York: Bantam Books, 1992), 171. All the quotations by Hawking printed in that book, which is hereafter cited as *A Reader's Companion*, have been reprinted in Stephen Hawking, *Black Holes and Baby Universes and Other Essays* (New York: Bantam Books, 1993); some of the quotations come from Stephen Hawking, *A Brief History of Time* (New York: Bantam Books, 1988).
15. Lubow, "Heart and Mind," 53; see also "Stephen Hawking," *Guardian*, April 19, 1993.
16. Yeo, "Genius, Method, and Morality," 258.
17. Stephen Greenblatt, *Renaissance Self-Fashioning: From More to Shakespeare* (Chicago: University of Chicago Press, 1980), 1. On Newton's self-fashioning, see Patricia Fara, *Newton: The Making of Genius* (New York: Columbia University Press, 2002); and Robert Iliffe, "'Is He Like Other Men? The Meaning of the *Principia Mathematica*, and the Author as Idol," in *Culture and Society in the Stuart Restoration: Literature, Drama, History*, ed. Gerald MacLean (Detroit, MI: Western University Press, 1995), 159–76.
18. Mario Biagioli, *Galileo Courtier: The Practice of Science in the Culture of Absolutism* (Chicago: University of Chicago Press, 1993), 5.
19. Talk given to the International Motor Neurone Disease Society (Zurich, 1987). See also Hawking, ed., *A Reader's Companion*, 4; and Hawking, *Black Holes and Baby Universes*, 1. The text of the talk was included with the August 1991 material and appears in *Black Holes and Baby Universes*.
20. It is interesting to see how the "Yes" at the beginning of the sentences and "[smiles]" give a sense of "being there" and "having" a real conversation. See Stephen Haw-

king, "Playboy Interview: Stephen Hawking—Candid Conversation," *Playboy*, April 1990, 64. See also "The Observer Profile: Stephen Hawking, Brief History of Genius," *Guardian*, September 10, 1995, 5: "It is perhaps one of those oddities of serendipity that 8 January 1942 was both the three-hundredth anniversary of the death of one of history's greatest intellectual figures, the Italian scientist Galileo Galilei, and the day Stephen William Hawking was born into a world torn apart by war and global strife. But, as Hawking himself points out, around two hundred thousand other babies were born that day, so maybe it is after all not such an amazing coincidence" (John Gribbin and Michael White, *Stephen Hawking: A Life in Science* [New York: Dutton, 1992], 5). See also Jane Hawking, *A Life with Hawking: Music to Move the Stars* (London: MacMillan, 1999), 201: "Galileo died on 8 January 1642, the year in which Newton was born and 300 years to the day before Stephen was born. Stephen adopted Galileo as his hero"; and 477: "Isaac Newton was born in 1642, the year in which Galileo died and 300 years before Stephen was born." See also George Johnson, "What a Physicist Finds Obscene," *New York Times*, February 16, 1997, 4: "Dr. Hawking has let it be known that he was born on the anniversary of Galileo's death (a coincidence of great astrological significance) and that he holds the very same chair at Cambridge University that Isaac Newton once did."

21. Hawking, ed., *A Reader's Companion*, 119.
22. Ibid., paraphrasing from 120–21.
23. As Greenblatt says, "Self-fashioning for such figures involves submission to an absolute power or authority situated at least partially outside the self, God, a sacred book, an institution such as church, court, colonial or military administration" (*Renaissance Self-Fashioning*, 9).
24. Hawking, ed., *A Reader's Companion*, 151–52.
25. Ibid., 92 (my emphasis). The first two sentences of this quotation effectively repeat sentences from Stephen Hawking, *A Brief History of Time* (New York: Bantam Books, 1988), 113: "However, one evening in November that year, shortly after the birth of my daughter, Lucy, I started to think about black holes as I was getting into bed. My disability makes this rather a slow process, so I had plenty of time." See also Hawking, *Black Holes and Baby Universes*, 18: "I was thinking about black holes as I got into bed one night in 1970, shortly after the birth of my daughter Lucy. Suddenly I realized that many of the techniques that Penrose and I had developed to prove singularities could be applied to black holes. In particular, the area of the event horizon, the boundary of the black hole, could not decrease with time. . . . I was so excited that I did not get much sleep that night." See also Tim Radford, "Book Thoughts: Master of the Universe," *Guardian*, December 30, 1988. According to Ferguson, citing Hawking, *A Brief History of Time*, 99, "There's an almost Olympian assuredness about touches of his writing. 'However, one evening, late in November that year, shortly after the birth of my daughter Lucy, I started to think about black holes as I was getting into bed. My disability makes this a slow process, so I had plenty of time,' and we are into a cerebration that ends with emissions from primordial black holes with masses greater than a fraction of a gram." See Kitty Ferguson, *Stephen Hawking:*

Quest for a Theory of Everything (New York: Bantam Books, 1992), 72; For Jane Hawking, "Stephen's night-time routine was a slow one, not only because of the physical constraints but also because his concentration was always directed elsewhere, usually on to a relativistic problem. One evening he took even longer that usual to get into bed; it was not until the next morning that I found out why. That night, while putting on his pyjamas and visualizing the geometry of black holes in his head, he had solved one of the major problems in black hole research." See Jane Hawking, *A Life with Hawking*, 176.

26. John Boslough, *Beyond the Black Hole: Stephen Hawking's Universe* (London: Collins, 1985), 63–64.
27. With an exception; this version is not in *Black Holes and Baby Universes*.
28. On the role of the narratives as repertoire constitutive of invention, see Mialet, *L'Entreprise créatrice*, and "Do Angels Have Bodies? Two Stories about Subjectivity in Science: The Cases of William X and Mr. H," *Social Studies of Science* 29, no. 4 (1999): 551–82
29. Joseph P. McEvoy and Oscar Zarate, *Introducing Stephen Hawking* (New York: Totem Books, 1995), 124 (my emphasis).
30. Hawking, *Brief History of Time*, 49; see also chap. 2, n. 5: "It would be easy . . . to be an astrophysicist because that is all in the mind. No physical ability is required" (Stephen Hawking, online at http://www.hawking.org.uk/index.php/about-stephen/questionsandanswers). In addition, Hawking tells of how, just after he had been told of his disease, when he was a young doctoral student looking for a subject for his thesis, he discovered the works of Penrose on the gravitational collapse of bodies.
31. Appleyard, "Master of the Universe," 29 (my emphasis).
32. Quoted from Hawking, "Playboy Interview," 68 (my emphasis).
33. Hawking, *Black Holes and Baby Universes*, 23.
34. "The Prospect of early death propelled him into a stupefying depression for two years, a period in which he spent little time on his research and a great deal of time in his room listening to classical music—mostly Wagner—and reading science fiction. He also began 'drinking a fair amount'" (John Boslough, *Stephen Hawking's Universe* (New York: Avon Books, 1985), 15.
35. Hawking, "Playboy Interview," 66–68.
36. Hawking, *Black Holes and Baby Universes*, 23; Ferguson, *Stephen Hawking*, 41: "He holed up miserably in his college rooms, but he insists, 'Reports in magazine articles that I drank heavily are an exaggeration. I felt somewhat of a tragic character. I took to listening to Wagner.'"
37. These two quotations are taken from Lisa Kremer, "The Smartest Person in the World Refuses to Be Trapped by Fate," *Morning News Tribune*, July 2, 1993; available online at http://weber.u.washington.edu/d27/doit/Press/hawking3.html.
38. See Inguun Moser and John Law, "Good Passages, Bad Passages," in *Actor Network Theory and After*, ed. Law and Hassard (Oxford: Blackwell, 1999), 199–220; and John Law and Inguun Moser, "Making Voices: New Media Technologies, Disabilities and Articulation," in *Digital Media Revisited: Theoretical and Conceptual Innovation in*

Digital Domain, ed. Gunnar Liestøl, Andrew Morrison, and Terje Rasmussen (Cambridge, MA: MIT Press, 2003) 491–520.

39. Both quotations are from Lubow, "Heart and Mind," 47.
40. Extracted from a longer unpublished piece in which Harry Collins compares the event with a religious ceremony. I thank Harry Collins for sharing his notes with me.
41. William Hicklin, interview by the author, January 31, 2007.
42. Ibid.
43. Ibid.
44. Ibid.
45. Ibid.
46. Gribbin wrote the science chapters (the even numbered chapters), and White wrote the life chapters (the odd ones). Gribbin first joined as a consultant and then became a co-author. Hawking was a postdoc at Cambridge when Gribbin was doing his PhD. They neither worked together nor were friends, but he knew Hawking and his work. One of the conditions on which he joined the project was that Gribbin would ask Hawking if he would mind. Hawking said no. Hawking said he was too busy to be involved, but he was happy that Gribbin would be able to interpret his work. "So we didn't have his active participation in the project, but he certainly didn't mind us doing it" (John Gribbin, interview by the author, September 16, 1997). White conducted many of the interviews with Hawking's colleagues and possibly his family, but they never interviewed Hawking in person. They were provided with some background material, a collection of essays that became part of *Black Holes and Baby Universes*, a pamphlet about his illness, and some other unpublished material. "He didn't vet the material; he just said get on with it." They used Kitty Ferguson's book. "She actually did have access—it was because he'd been talking to her that he didn't talk to us, because he didn't have time to do it again. And, so we were just behind her in that sense" (ibid.).
47. I used this article as well; see http://math.ucr.edu/home/baez/README.html.
48. William Hicklin, interview by the author, January 31, 2007.
49. Ibid.
50. Ibid.
51. Ibid.
52. Ibid.
53. Ibid. Though we'll see that Hawking is, in fact, very active in the construction of his image.
54. As Erving Goffman puts it, "Personal identity, then, has to do with the assumption that the individual can be differentiated from all others and that around this means of differentiation a single continuous record of social facts can be attached, entangled, like candy floss, becoming then the sticky substance to which still other biographical facts can be attached" (*Stigma: Notes on the Management of Spoiled Identity* [Englewood Cliffs, NJ: Prentice-Hall, 1963], 57).
55. William Hicklin, interview by the author, January 31, 2007.

56. Ibid.
57. This recalls what Albert Robillard, who has ALS and who is a professor of sociology, says: "These assistants are my legs, arms, and whole bodies. They are also my trusted colleagues. I use them, for example, to enter my lip-signed letters into the computer. But they also go to the library, pick up the mail, get video equipment, tape scenes about town, turn pages and arrange documents, sign my name to forms, go with me to committee and classes to translate, speak for me on the phone, read my papers before professional meetings, keep and maintain files, translate for me when I have visitors, get me out of the car and into and out of my office chair, translate for me when I'm counseling students, suction my mouth, remind me of things to do, and carry out personal tasks. Beyond all this, they are also people I confide in and who will tell me how to do things better. These assistants are more than employees. There is also a constant dialogue going between the students and myself about their lives and my life (even if much of the talk is rank gossip)." See Albert Robillard, *Meaning of a Disability: The Lived Experience of Paralysis* (Philadelphia, PA: Temple University Press, 1999), 162.
58. Christophe, interview by the author, July 8, 2005.
59. William Hicklin, interview by the author, January 31, 2007.
60. Ibid.
61. Ibid.
62. Ibid.
63. Christophe, interview by the author, July 8, 2005.
64. William Hicklin, interview by the author, January 31, 2007.
65. Ibid.
66. Sam, interview by the author, June 12, 2007.
67. As I was asking Christophe why Hawking didn't retrieve the quotation where he compares himself to Galileo from his computer, he looked surprised and told me that he didn't know that Hawking had used this quote before. He added that maybe as an anthropologist I know everything Hawking wrote, but it doesn't mean that Hawking memorizes everything he writes.
68. Christophe, interview by the author, July 8, 2005.
69. William Hicklin, interview by the author, January 31, 2007.
70. Ibid.
71. In a sense, they are reconstructing a person who is already a construction.
72. William Hicklin, interview by the author, January 31, 2007.
73. Errol Morris, quoted in James Delingpole, "Limelight," *Evening Standard*, June 27, 1990, cited in Michael White and John Gribbin, *Stephen Hawking*, 283.
74. William Hicklin, interview by the author, January 31, 2007.
75. Ibid.
76. Ibid.
77. Ibid.
78. Ibid.
79. Ibid.; my emphasis.

80. They are obsessed with the idea of being true to the original in the same way as Errol Morris, who reconstructed an exact replica of Hawking's wheelchair from scratch in his movie *A Brief History of Time*.
81. William Hicklin, interview by the author, January 31, 2007.
82. Ibid.
83. Hicklin seems to think that this is, in part, a technical problem: one cannot do this with Hawking's computer. His assistant told me that Hawking wanted to do it on his own.
84. William Hicklin, interview by the author, January 31, 2007.
85. Christophe, interview by the author, July 8, 2005.
86. I told Christophe that I'd love to see Hawking working. He responded, "You just saw him working."
87. GA Sam talking about EZ Keys and Equalizer on June 12, 2007.
88. Written version given to me by William Hicklin, 2005. I didn't change the orthographical mistakes or typos, as they indicate the presence of the author and make the difference between the "similar" statements visible.
89. Another of these commonplace descriptions deals with Wheeler's naming of the black hole in 1969: "The term *black hole* was coined by John Wheeler in 1969, for what had previously been known as a collapsed object, or frozen star. It was a brilliant move. It not only caught people's imagination, but it focused attention on the hole left behind, which was independent of the details of the body that collapsed to produce it" (Hawking, *Black Holes and Baby Universes*, 116).
90. "I don't care much for equations myself," Hawking says now. "This is partly because it is difficult for me to write them down but mainly because I don't have an intuitive feeling for equations. Instead, I think in pictorial terms" (William H. Cropper, *Great Physicists: The Life and Times of Leading Physicists from Galileo to Hawking* [Oxford: Oxford University Press, 2001], 453; quoted by Hawking in "A Brief History of a *Brief History*," *Popular Science*, August, 1989, 70). Kitty Ferguson writes, "He claims he's not overly fond of equations himself, in spite of the fact that people compare his ability to handle them in his head with Mozart's mentally composing a whole symphony. It's not easy for him to write equations, and he says he has no intuitive feelings for them. He likes to think instead in pictures" (*Stephen Hawking*, 130). John Noble Wilford writes, "'I avoid problems with a lot of equations or translate them into problems of geometry,' Hawking said recently. 'I can then picture them in my mind'" ("Scientist at Work: Stephen Hawking—Sailing a Wheelchair to the End of Time," *New York Times*, March 24, 1998).
91. "As a young boy, I was the despair of my parents because I was always taking things apart to see what made them tick. Of course I generally couldn't get them back together again, but I felt that if I understood how something worked, then in a sense I was the master of it" (Stephen Hawking, quoted in the foreword to David Filkin, *Stephen Hawking's Universe: The Cosmos Explained* [New York: Basic Books, 1998], xiii); "I was always very interested in how things operated and used to take them apart to see how they worked, but I was not so good at putting them back together again. . . . But from the age of thirteen or fourteen, I knew I wanted to do research

in physics because it was the most fundamental science" (Hawking, *Black Holes and Baby Universes*, 11); "J: Then you had made up your mind early on what you wanted to do? SH: Yes. From the age of twelve, I had wanted to be a scientist. And cosmology seemed the most fundamental science" (Hawking, "Playboy Interview, " 66).

92. William Hicklin, interview by the author, January 31, 2007 (my emphasis).
93. Ibid.
94. Ibid.
95. Ibid.
96. Ibid.
97. As we will see in the last chapter, Hawking uses the same expression when he worries about being taken for a figurehead.
98. William Hicklin, interview by the author, January 31, 2007.
99. Ibid.
100. Ibid.
101. Ibid.
102. Lubow, "Heart and Mind," 46.
103. His humor can be seen as the product of the system of which he is part. Indeed, as Sacks notices, this is the only way of participating in a conversation if one cannot interact (i.e., cannot have a back-and-forth discussion); see Harvey Sacks, *Lectures on Conversation*, vol. 1 (Malden, MA: Blackwell, 1992).
104. William Hicklin, interview by the author, January 31, 2007.
105. Ibid.
106. Ibid.
107. Behind the scenes: Hicklin tells me that he thinks Hawking didn't like the critiques at the end of the film but that Christophe found them quite fair; in any case he seemed to have no problem with the way in which they explained science. Christophe later tells me that he thinks Hawking liked the movie but that he personally didn't like the way the science was explained.

Chapter Five

1. In a way, what I try to question in this context is a certain ideology claiming the primacy of the field and the immediacy of experience that accompanies it, the idea that the *mise-en-présence* with the other would give the analyst better access to the true nature of things. I also engage with the idea, already called into question by Goffman, that it is possible to get rid of the different layers that constitute a person so as to reach "the true self." See Erving Goffman, *The Presentation of Self in Everyday Life* (Garden City, NY: Doubleday, 1959).
2. This question could be linked to certain reflections about the relationship between historiography and anthropology. See Paul Ricoeur, "What Is a Text? Explanation and Understanding," and "The Model of the Text: Meaningful Action Considered as a Text," in *Hermeneutics and the Human Sciences: Essays on Language, Action and Interpretation*, trans. and ed. John B. Thompson (Cambridge: Cambridge University Press, 1981), 145–64 and 197–221, respectively. See also chapter 6.

3. For a critique of the ideology that claims that representations produced by anthropologists (e.g., keeping good field notes, making accurate maps, writing up results) are transparent, see James Clifford, "Introduction: Partial Truths," in *Writing Culture: The Poetics and Politics of Ethnography*, ed. James Clifford and Georges E. Marcus (Berkeley and Los Angeles: University of California Press, 1986), 1–26.
4. See, for example, Isabelle Stengers, who proposes that we give the subjects of the human sciences the power of putting into question the tools of the analyst. *Pour en finir avec la tolérance*, vol. 7 of *Cosmopolitiques* (Paris: Editions La Découverte/Les Empêcheurs de penser en rond, 1997).
5. See Hélène Mialet, "Is the End in Sight for the Lucasian Chair? Stephen Hawking as Millennium Professor," in *From Newton to Hawking: A History of Cambridge University's Lucasian Professorship of Mathematics*, ed. Kevin Knox and Richard Noakes (Cambridge: Cambridge University Press, 2003), 425–59.
6. It would probably be interesting to analyze the relation between mystification and fear.
7. Gene Stone, quoted in Arthur Lubow, "Heart and Mind," *Vanity Fair*, June 1992, 44–53. Stone was responsible for organizing the transcripts from the film interviews into Stephen Hawking's *A Brief History of Time: A Reader's Companion*, prepared by Gene Stone (New York: Bantam Books, 1992).
8. The commutator amplifies and makes audible the curling of his fingers.
9. Professor Hawking's responses to my questions have been transcribed exactly as they were printed out; I have kept his spelling without making any changes or corrections. See fig. 6 for a facsimile of the printout.
10. See Kip Thorne, *Black Holes and Time Warps: Einstein's Outrageous Legacy* (New York: W. W. Norton, 1994), 481–82.
11. Some machines are very noisy.
12. Except if we communicated by e-mail. See Harold Garfinkel, *Studies in Ethnomethodology* (Cambridge: Cambridge University Press, 1967).
13. Like Garfinkel's famous breaching experiment that involves disturbing routines, Hawking's disruption reveals something about the necessary requirements of fluent social interactions/conversation and about the disruption itself. It also tells us that 'who' Hawking *is* reflects differences in the microsociology of face-to-face relations with him, especially with regard to the importance of familiarity in this process of communication. See for example, David Goode, *A World without Words: The Social Construction of Children Born Deaf and Blind* (Philadelphia, PA: Temple University Press, 1994).
14. The disappearance of the role of assistants and instruments in the construction of scientific facts is related partly to the scientist's rhetoric when he constructs himself as the sole author of facts produced and partly to scientific practice itself, which aims to objectify facts. Yet that which is normally hidden in a single corporal envelope when it concerns a scientist endowed with an active body becomes, with Hawking, exteriorized because of his disability. We can thus see, with him, that which is normally visible but intentionally hidden, as well as that which is normally present but constitutionally hidden.

15. After the present text was completed, I discovered a fascinating book written by Albert Robillard (*Meaning of a Disability: The Lived Experience of Paralysis* [Philadelphia, PA: Temple University Press, 1999]). Robillard has ALS and describes—with the tools of ethnomethodology—the implication of his disease for interactions in daily life. I found, from the point of view of one who can't read an interaction, the same things he found through his difficulty with everyday interactions and from his feelings of isolation: the role of mutual gaze, what he calls "real time" in conversation, and in general the use of the body and voice to create and maintain participation in a social setting.
16. Harvey Sacks, Emanuel A. Schegloff, and Gail Jefferson, "A Simplest Systematics for the Organization of Turn-Taking for Conversation," *Language* 50, no. 4 (1974): 696–735.
17. The question "Why you?" can be interpreted as being a trick to give him time to respond to my question. This bridges the gap occurring during the conversation, just as his assistants do when, during conferences, they explain the workings of Professor Hawking's computer while he composes his answers to questions. We can also say here that in a certain sense he does not respond like a machine. Nevertheless, Hawking would not be able to pass the Turing Test; because of his slowness at answering questions, someone who is not in his presence might have the feeling that he or she is communicating with a machine rather than with a human being.
18. Sacks, Schegloff, and Jefferson, "Simplest Systematics," 714.
19. We see also different ways in which his silence is significant for pragmatic reasons. This silence encourages people to project and create meanings.
20. Sacks, Schegloff, and Jefferson, "Simplest Systematics," 727–28. See also Charles Goodwin, *Conversational Organization: Interaction between Speakers and Hearers* (New York: Academic Press, 1981), and "Co-Constructing Meaning in Conversations with an Aphasic Man," *Research on Language and Social Interaction* 28, no. 3 (1995): 233–60.
21. It is important to note that researchers who focus on gestural language reveal its unnoticed, because hidden, presence by "tricks of language." In my case, its presence is noticed because it is absent. Formal language is thus totally detached from gestural language.
22. Mary Rose Barral, *Merleau-Ponty: The Role of the Body-Subject in Interpersonal Relations* (Pittsburgh, PA: Duquesne University Press, 1965), 92–93.
23. Goffman, *Presentation of Self*.
24. See, for example, Adam Kendon, "Some Functions of Gaze Direction in Social Interaction," *Acta Psychologica* 26, no. 1 (1967): 22–63.
25. "According to M. Gidding, [says Gabriel Tarde] when two men meet, the conversation they have is just a complement of their reciprocal gaze by which they explore and try to know if they belong to the same social group." Or again, "We cherish, he says, the illusion that makes us believe that we speak because we care about the things we are talking about, in the same way we cherish this illusion, the sweetest of all, the belief in art for the sake of art. The truth is that any expression, by the commoner and by the artist, and every communication, from the accidental

conversation upon meeting for the first time to the profound intimacy of true love, have their origin in the elementary passion of knowing one another and to make known mutually to one another, to define the consciousness of the species." Tarde adds, "Subsequently, conversations tend, as a matter of course, to give birth to and to accentuate, to extend and to deepen *the consciousness of species*, not only to define it. It is not a question of making visible its limits, but to push them endlessly." See Gabriel Tarde, *L'opinion et la foule* (Paris: Presses Universitaires de France, 1989), 62 (my translation).

26. Paul Ricoeur, *Hermeneutics and the Human Sciences: Essays on Language, Action and Interpretation*, trans. and ed. John B. Thompson (Cambridge: Cambridge University Press, 1981), 146–47.
27. Ibid., 200.
28. See ibid., 199. On the contrary, Ricoeur argues that what writing fixes is not the event of speaking but the "said" of speaking.
29. See Stefan Hirschauer, "The Manufacture of Bodies in Surgery," *Social Studies of Science* 21 (May 1991): 279–319.
30. We can say that we all have a more or less routinized environment around us. Nevertheless, the able-bodied seem able to navigate more or less freely in this environment. Hawking never manages to escape from this little microenvironment, this extended body that I have just described.
31. In this sense, there are obvious parallels between Hirschauer's article on surgery ("The Manufacture of Bodies in Surgery"), Edwin Hutchins on the functioning of a cockpit ("How a Cockpit Remembers Its Speeds," *Cognitive Science* 19 [July–September 1985]: 265–88), and this chapter on Hawking.
32. Michael Polanyi, *Knowing and Being*, ed. Marjorie Grene (Chicago: University of Chicago Press, 1969), 148.
33. Walt Woltosz, interview by the author, December 22, 2000.
34. Michael Polanyi, *The Study of Man: The Lindsay Memorial Lectures* (Chicago: University of Chicago Press, 1959), 31.
35. Drew Leder, *The Absent Body* (Chicago: University of Chicago Press, 1990), 179, n. 70.
36. Leder, *Absent Body*, 34. See Polanyi, *Knowing and Being*, 145; and Maurice Merleau-Ponty, *Phenomenology of Perception*, trans. Colin Smith (London: Routledge and Kegan Paul, 1962), 143; or the example of the handheld tool discussed in Martin Heidegger, *Being and Time*, trans. John Macquarrie and Edward Robinson (New York: Harper & Row, 1962), 95–107. Don Ihde describes a similar phenomenon: "I may describe these relations as *embodiment relations*, relations in which the machine displays some kind of partial transparency so that it itself does not become objectified or thematic, but is taken into my experiencing of what is other in the World"; see his *Technics and Praxis* (Dordrecht, Holland: D. Reidel, 1979), 8, quoted in Leder, *Absent Body*, 179, n. 69 (see also n. 44).
37. Stephen Hawking, quoted in Colker, "Giving a Voice to the Voiceless," *Los Angeles Times*, May 13, 1997, 8.
38. Michael Polanyi, *Study of Man*, 31.

39. Michael Polanyi, *The Tacit Dimension* (Garden City, NY: Doubleday, 1966), 15.
40. It would be interesting to explore the relationship between memory and automatism. This echoes what David, who is quadriplegic, said to me during an interview conducted in December 2000: "I don't experience myself as a disabled individual all the time. I am aware of it, but it is not the focus of my life. At some point in time disability stopped being the main thing in my life and started just being a major inconvenience, which pops its head up every once in a while, especially when I need something that I can't get."
41. This point agrees with what Leder says: "I do not notice my body, but neither do I, for the most part, notice the bed on which I sleep, the clothes I wear, the chair on which I sit down to breakfast, the car I drive to work. I live in bodies beyond bodies, clothes, furniture, room, house, city, recapitulating in ever-expanding circles aspects of my corporeality. As such, it is not simply my surface organs that disappear but entire regions of the world with which I dwell in intimacy" (Leder, *Absent Body*, 35).
42. "This transparency is never complete; as Ihde points out, there is always an 'echo focus' (in *Technics and Praxis*, 7), a subliminal awareness of the instrument at my body boundary. Moreover, especially at times of malfunction, the tool can become thematically central" (Leder, *Absent Body*, 179, n. 69). "When incorporating a tool, one's concurrent incorporation by it may have a series of unanticipated results. As Ihde notes, every instrument imposes an 'amplification-reduction' structure on one's natural capacities (*Technics and Praxis*, 9–10, 21–26). The telephone, while allowing communication through its amplification of voice, in other ways yields a reduced encounter devoid of direct sight and touch. As a result of such transformations, each technology provides a 'telic inclination' toward certain modified modes of interaction (42–44)." (Leder, *Absent Body*, 181, n. 72; see also a similar argument in Don Ihde, *Bodies in Technology* [Minneapolis: University of Minnesota Press, 2002], chap. 1); hence, the problem this implies in my mode of communication with Hawking.
43. René Descartes, *The "Meditations" and Selections from the "Principles" of Philosophy*, trans. John Veitch (Chicago: Open Court, 1913), 94.
44. Barral, *Merleau-Ponty*, 94. Also, according to Polanyi, "We must realize then also that our own body has a special place in the universe: we never attend to our body as an object in itself" (*Study of Man*, 31); or, "Dwelling in our body clearly enables us to attend *from* it to things outside, while an external observer will tend to look *at* things happening in the body, seeing it as an object or as a machine" (*Knowing and Being*, 148).
45. Merleau-Ponty, *Phenomenology of Perception*, 92.
46. Ibid., 150.
47. Maurice Merleau-Ponty, *The Structure of Behavior*, trans. Alden L. Fisher (Boston: Beacon Press, 1963), 213.
48. Louis-Marie Morfaux, *Vocabulaire de la philosophie et des sciences humaines* (Paris: Armand Colin, 1980), 67, s.v. *corps*; translation mine.
49. He makes visible the dysfunctions of his flesh-and-blood body but also all his normal needs, such as the need to eat and drink.

50. Ernst H. Kantorowicz, *The King's Two Bodies: A Study in Medieval Political Theology* (Princeton, NJ: Princeton University Press, 1957), 13. Kantorowicz adds, "Interesting, however, is the fact that this 'incarnation' of the body politic in a king of flesh not only does away with the human imperfections of the body natural, but conveys 'immortality' to the individual king as King, that is, with regard to his superbody" (13). We also witness a dissociation between corporeal, mental, and social phenomena, such as the construction of the self, "normally" inscribed in or displayed *simultaneously* by "one" body.
51. The text handed to me is thus doubly disembodied. On one hand, its conditions of production have been erased, but this aporia demands that the text, if it is to be coherent, refers back to this absence—that is, to the conditions that made it possible. On the other hand, as it is about Stephen Hawking, the Lucasian Professor of Mathematics who falls into the long lineage of his ancestors, his peculiar individual conditions are no longer significant so long as he produces scientific theories.
52. I used this text both to write the present chapter and, previously, "Is the End in Sight for the Lucasian Chair?"

Chapter Six

1. As shown in the previous chapter, these extended bodies are naturalized and participate in the construction of Hawking's sacred body.
2. Similar to Hawking's computer, which, at its user's discretion, can save articles, lectures, and interviews while leaving no trace of "informal" interactions, we see a collective at work.
3. On the history of the emergence of the concept of the archival body, see Mario Wimmer, *Archivkorper: Eine Geschichte historischer Einbildungskraft* (Constance: Konstanz University Press, 2012).
4. Head of the Moore Library, interview by the author, August 9, 2005.
5. They have a technical connection, insofar as Hawking can be seen as an emblem for state-of-the-art technology.
6. Approval for construction of a new library in Cambridge, online at http://www.maths.cam.ac.uk/Friends/newsletters/news5/.
7. Annelise Riles, ed., *Documents, Artifacts of Knowledge* (Ann Arbor: University of Michigan Press, 2006) is an exception; some of the authors of this edited collection are (for the most part) anthropologists studying documentary practices in different ethnographic contexts, but none of them focuses specifically on archivists' work in a library.
8. On writing ethnography and writing history, see James Clifford and George Marcus, eds., *Writing Culture: The Poetics and Politics of Ethnography* (Berkeley and Los Angeles: University of California Press, 1986); Clifford Geertz, *Works and Lives: The Anthropologist as Author* (Stanford, CA: Stanford University Press, 1988); Hayden White, *Metahistory: The Historical Imagination in Nineteenth-Century Europe* (Baltimore, MD: Johns Hopkins University Press, 1973); Dominick LaCapra,

Writing History, Writing Trauma (Baltimore, MD: Johns Hopkins University Press, 2001).

9. Head of the Moore Library, interview by the author, August 9, 2005.
10. Curator of scientific manuscripts, interview by the author, June 20, 2007.
11. There is a form of selection. What he is not interested in becomes part of the archives.
12. Librarian's assistant, interview by the author, August 9, 2005.
13. Head of the Moore Library, interview by the author, June 18, 2007.
14. Head of the Moore Library, interview by the author, August 9, 2005.
15. Head of the Moore Library, interview by the author, March 30, 2007.
16. I take the expression *paper organism* from Mario Wimmer, "Die kalte Sprache des Lebendigen: Zu den Anfängen der Archivberufssprache," in *Sprachvollzug im Amt Kommunikation und Verwaltung im Europa des 19 und 20 Jahrhunderts*, ed. Peter Becker (Bielefeld, Germany: Transcript, 2011).
17. Head of manuscripts, interview by the author, May 18, 2007.
18. According to the head of manuscripts, fan letters sent to Hawking are not part of the archives.
19. Head of the Moore Library, interview by the author, June 18, 2007.
20. Curator of scientific manuscripts, interview by the author, June 20, 2007. As for Hawking's personal assistant, she told me, "If something happened to Stephen Hawking, my dream would be to seal his office and to keep everything." When I mentioned this to the head of manuscripts, he responded, "Well, yes, I mean, it needs to be kept, . . . in order to be appraised, and . . . select material, and so on. . . . She might mean as a sort of shrine, . . . but then, of course, nobody can consult it, if it's a shrine. I mean, that's something quite different, isn't it? . . . I mean, what we're interested in [in] a library is making material available for research, which is a rather different exercise" (Head of manuscripts, interview by the author, June 20, 2007).
21. Departmental archives will be kept by the department and administered differently by the university.
22. Curator of scientific manuscripts, interview by the author, May 16, 2007.
23. Head of the Moore Library, interview by the author, June 18, 2007.
24. Head of the Moore Library, interviewed by the author, August 9, 2005.
25. Head of manuscripts, interview by the author, May 18, 2007.
26. Head of manuscripts, interview by the author, June 20, 2007.
27. Ibid.
28. Curator of scientific manuscripts, interview by the author, June 20, 2007.
29. Head of the Moore Library, interviewed by the author, August 9, 2005.
30. Ibid.
31. Head of manuscripts, interview by the author, May 18, 2007.
32. Head of the Moore Library, interview by the author, August 9, 2005.
33. Ibid.
34. Michael Lynch, "Archives in Formation: Privileged Spaces, Popular Archives and Paper Trails," *History of the Human Sciences* 12, no. 2 (1999): 69.

35. Head of the Moore Library, interview by the author, August 9, 2005.
36. Head of the Moore Library, interview by author, March 30, 2007.
37. Head of the Moore Library, interview by the author, August 9, 2005.
38. Ibid.
39. Head of the Moore Library, interview by the author, March 30, 2007. Paradoxically, we will see that most of the material will be stored at the UL.
40. Ibid. It is unusual to archive objects, though "we have all these spaces where we could exhibit things upstairs; and we could use these rooms. . . . We could put something which is a bit more liable for theft to be locked up a bit more securely."
41. Similarly, one could imagine a scenario where Hawking could decide that the material they have would be made accessible during his lifetime.
42. Head of the Moore Library, interview by the author, August 9, 2005.
43. Head of the Moore Library, interview by the author, March 30, 2007.
44. Librarian's assistant, interview by the author, August 9, 2005.
45. Ibid.
46. Ibid.
47. Ibid.
48. Head of the Moore Library, interview by the author, June 18, 2007.
49. Letter from Anthony Edwards, University Library. Classmark MS. Add.9222.
50. Head of manuscripts, interview by the author, May 18, 2007. There are probably differences between this version and the published one.
51. They have various drafts of *A Brief History of Time* as well as lectures published from that book; they also have the draft of *Black Holes and Baby Universes*.
52. Librarian's assistant, interview by the author, August 9, 2005.
53. In the same way, the head of the Moore says, "We have a gown, but we are not quite sure where it belongs and what institution it was [from which] he brought it back" (interview by the author, August 9, 2005).
54. Librarian's assistant, interview by the author, August 9, 2005.
55. Ibid.
56. Ibid. Paragraphs must be reintroduced in speeches as well.
57. Michel Foucault, "What Is an Author?" in *Language, Counter-Memory, Practice*, trans. Donald F. Bouchard and Sherry Simon (Ithaca, NY: Cornell University Press, 1977), 127.
58. Head of the Moore Library, interview by the author, August 9, 2005.
59. Head of the Moore Library, interview by the author, June 18, 2007.
60. Librarian's assistant, interview by the author, August 9, 2005.
61. See for example, Hegel or, more recently, Bourdieu, Derrida, or Foucault. For example, Derrida's *Archive Fever* originated as a lecture delivered in 1994 and was first published in French in 1995 as *Mal d'archive: Une impression freudienne*. It was translated and published the same year in the American journal *Diacritics* and as a separate English-language monograph in 1996.
62. Head of the Moore Library, interview by the author, June 18, 2007.
63. Lilly Koltun, "The Promise and the Threat of Digital Options in an Archival Age," *Archivaria* 47 (Spring 1999): 119. On the implications of digital media for the tradi-

tional notion of an archive, see also Rudi Laermans and Pascal Gielen, "The Archive of the Digital An-archive," *Image and Narrative: An Online Magazine of the Visual Narrative*, no. 17 (April 2007): n.p. Online at www.imageandnarrative.be/.

64. Head of manuscripts, interview by the author, May 18, 2007.
65. Bruno Latour, "Visualization and Cognition: Thinking with Eyes and Hands," in *Knowledge and Society: Studies in the Sociology of Culture Past and Present*, vol. 6 (1986): 1–40.
66. Bruno Latour, *Science in Action: How to Follow Scientists and Engineers Through Society* (Cambridge, MA: Harvard University Press, 1987).
67. Head of the Moore Library, interview by the author, June 18, 2007. What distinguishes the text from the performance is the presence of the man.
68. This recalls Walter Benjamin's "The Work of Art in the Age of Mechanical Reproduction" (in his *Illuminations* [New York: Schocken Books, 1968]), though the argument here is to show that the work of permanent translation through different media maintains aura. See also Bruno Latour and Adam Lowe, "The Migration of the Aura; or, How to Explore the Original Through Its Facsimiles," in *Switching Codes*, ed. Thomas Bartscherer (Chicago: University of Chicago Press, 2010).
69. Mike Featherstone, "Archive," *Theory, Culture and Society* 23, nos. 2–3 (May 2006): 593–94.
70. Librarian's assistant, interview by the author, August 9, 2005.
71. Ibid.
72. Ibid. It's easier to talk to his assistant than to him; however, the head of manuscripts, "who is a real archivist," mentioned that if they don't know, they don't; they are not magicians.
73. Librarian's assistant, interview by the author, August 9, 2005.
74. She thinks the archivists at the University Library will do it, though she doesn't know what priority it has for them.
75. She is probably referring to the collection of materials and their coherences as well.
76. Or she creates differences.
77. Librarian's assistant, interview by the author, August 9, 2005.
78. Ibid.
79. Ibid.
80. Head of the Moore Library, interview by the author, August 9, 2005.
81. Head of the Moore Library, interview by the author, June 18, 2007.
82. Librarian's assistant, interview by the author, August 9, 2005.
83. Ibid. See also William Hicklin, who said in my interview with him (January 31, 2007) that his desire was "to know him."
84. Interestingly, the archivist sees her job as a way of presenting the material to the public more than telling a story. As Carolyn Steedman says, "Historians read for what is not there: the silences and the absences of the documents always speak to us" ("Something She Called a Fever: Michelet, Derrida and Dust," *American Historical Review* 106, no. 4 [2001]: 1177).
85. Librarian's assistant, interview by the author, August 9, 2005. The assistant creates

a distinction between the Hawking who will be presented in the exhibition for the general public and the Hawking that will be reconstructed by the historians.

86. Curator of scientific manuscripts, interview by the author, May 16, 2007.
87. Ibid.
88. Curator of scientific manuscripts, interview by the author, June 20, 2007.
89. On the ways on which the intentions of voters were read through their ballots, see also Michael Lynch, Stephen Hilgartner, and Carin Berkowitz, "Voting Machinery, Counting and Public Proofs in the 2000 US Presidential Election," in *Making Things Public: Atmospheres of Democracy*, ed. Bruno Latour and Peter Wiebel (Cambridge, MA: MIT Press, 2005): 814–28.
90. Curator of scientific manuscripts, interview by the author, June 20, 2007.
91. Head of the Moore Library, interview by the author, June 18, 2007.
92. There is again an interesting parallel between Hawking's extended body and the archival records as "organic," that is, "developing naturally from their relation to agency or entity. The value of individual records depends directly on their relationship with those surrounding them. The pieces lose relevance if separated from the whole. The published materials in libraries are created independently, enabling a library collection to remain relevant even when materials are removed or checked-out" (Sara Schmidt, *Split Personalities: A Librarian in the Archives*, paper presented at the Society of American Archivists, Austin, TX, August 2009).
93. Curator of scientific manuscripts, interview by the author, June 20, 2007.
94. Curator of scientific manuscripts, interview by the author, May 16, 2007.
95. Curator of scientific manuscripts, interview by the author, June 20, 2007.
96. Ibid. For a very different interpretation, see Andrew Warwick, "Exercising the Student Body," in *Science Incarnate*, ed. Lawrence and Shapin.
97. Ibid.
98. See Steven Shapin, *A Social History of Truth: Civility and Science in Seventeenth-Century England* (Chicago: University of Chicago Press, 1994).
99. The curator of scientific manuscripts indicates that this new medium will affect their way of working. But if it affects the practices and content of their discipline, he doesn't seem to see why Hawking's reliance on the computer (because of his illness) should affect the content of his work (interview by the author, June 20, 2007).
100. Ibid. This problem with electronic media predates the widespread use of computer technology. For example, as Michael Lynch has pointed out to me in this regard, the archive of Harold Garfinkel contains important recordings (e.g., conversations with Goffman and others in the 1950s) on wire-recording devices. It would be very expensive to digitize these. One wonders to what extent—and how—Hawking's voice could be digitized.
101. Curator of scientific manuscripts, interview by the author, June 20, 2007.
102. See Latour and Lowe, "Migration of the Aura."
103. One can make a parallel with Hawking saying yes to papers scanned under his eyes.
104. Curator of scientific manuscripts, interview by the author, June 20, 2007.
105. Ibid.

106. Head of the Moore Library, interview by the author, August 9, 2005.
107. Ibid. We have seen that being close to or far from Hawking makes a difference in the way he is perceived; what then does it change to be in the physical presence of his relics or not?
108. Head of the Moore Library, interview by the author, June 18, 2007.
109. Ibid.
110. Ibid.
111. Michael Lynch and David Bogen, *The Spectacle of History: Speech, Text and Memory at the Iran-Contra Hearings* (Durham, NC: Duke University Press, 1996), 51.
112. "This would require access to a range of material which may well not make it into a freely available digital 'exhibition.' I imagine if anyone wanted to do this, they'd probably still have to come in person" (head of the Moore Library, e-mail message to author, February 16, 2010).
113. Head of the Moore Library, interview by the author, June 18, 2007.
114. Ibid.
115. Ibid.
116. For example, in 2009 there were several exhibitions in separate university museums—Zoology, Earth Sciences, the Fitzwilliams, and the University Library—to celebrate Darwin's two hundredth anniversary. In contrast, the future digital Hawking Archive exhibition could become a single university museum.
117. We'll be able to read texts by him and about him at the same time. The distinction between primary and secondary sources seems elided. On a similar situation, see Bruno Latour, "A Textbook Case Revisited: Knowledge as a Mode of Existence," in *The Handbook of Science and Technology Studies*, 3rd ed., ed. Edward J. Hackett, Olga Amsterdamska, Michael Lynch, and Judy Wajcman (Cambridge: MIT Press, 2007), 83–112.
118. Koltun, "Promise and the Threat of Digital Options," 133.
119. However, though it might well become virtual, the exhibition will also stay on the first floor of the Moore as an attraction for the general public and as a source of information for researchers. Moreover, the Center for Mathematical Sciences has become one of the architectural highlights of Cambridge.
120. Rudi Laermans and Pascal Gielen comment on the work of Boris Groys. They say that he "stresses that every kind of archive has a 'carrier medium,' such as paper or film, a computer or a computer network. The archive consists of signs or ensembles of signs, and the signs imply the existence of material carriers. It seems logical to include the carriers in the archive, yet Groys decisively argues against this view: 'Books are not part of the archive, but texts are; canvasses are not, but paintings are; video accessories are not, but moving images are. [. . .] The carrier of the archive does not belong to the archive, in that it carries the sign of the archive, without itself being a sign of the archive. [. . .] The sign carriers remain hidden behind the signs they carry. The archival carrier is fundamentally removed from the observer's view. The observer sees only the media surface of the archive; one can only guess at the media carriers'" (Laermans and Gielen, "The Archive of the Digital An-archive," 5).

121. See for example, Michael Hagner, *Cerveaux des génies*, tr. Olivier Mannoni (Paris: Éditions de la Maison des Sciences de l'Homme, 2008).
122. The archives can be seen, to use Mike Featherstone's expression, "as prosthetic memory devices for the re-constitution of an identity" (Featherstone, "Archive," 594). Hawking's computer literally materializes this expression.
123. Sam, interview by the author, June 12, 2007.
124. In Kevin Knox and Richard Noakes, eds., *From Newton to Hawking: A History of Cambridge University's Lucasian Professors of Mathematics* (Cambridge: Cambridge University Press, 2003). I was responsible for the chapter on Hawking, in which I used some of the information I gathered during my first interview with him; see chapter 5 above.
125. Harriet Bradley, "The Seductions of the Archive: Voices Lost and Found," *History of the Human Sciences* 12, no. 2 (1999): 119.

Chapter Seven

1. Gilbert Ryle, "The Thinking of Thoughts: What Is 'Le Penseur' Doing?" (Chapter 37 of Gilbert Ryle's collected papers (1971), online at http://web.utk.edu/~wverplan/ryle.html.
2. See her Web site: http://www.eveshepherd.com/commissions/hawking.shtml.
3. This is what I was told, but it doesn't seem to be the case.
4. The Gordon Moore library where the Hawking Archive are kept is at the opposite side of the statue.
5. Against a form of determinism, see Bruno Latour, "Which Politics for Which Artifacts" (*Domus*, June 2004, 50–51): "Haussmannian architects had found a way to keep servants and bourgeois apart, but they never anticipated relegating students to the back stairs. Technology, in other words, has its own intent and import which makes the best (or the worst of intent) drift away. This is why it's always difficult to do reverse engineering and to read the intention of a designer out of a design."
6. Roland Barthes, *Mythologies* (New York: Hill and Wang, 1987), 92.
7. Ibid., 93.
8. It is unseen except in the film *The Hawking Paradox*, where one sees a student writing long equations, but once again it was simply a scene set up for the camera and the public.
9. We see a similar phenomenon in the way Hicklin made the documentary. Remember that Hicklin mentioned that because Hawking can't move, he had to make the camera move around him.
10. This is the case for the statue as well.
11. To paraphrase Kantorowicz, the Lucasian Chair, if it is a "corporation by succession," is a "fiction which makes us think of the witches in Shakespeare's *Macbeth* ... who conjure up that uncanny ghostly procession of Macbeth's predecessor kings whose last one bears the 'glass' showing the long file of successors." See Ernst Kantorowicz, *The King's Two Bodies* (Princeton, NJ: Princeton University Press, 1957), 387.

12. The collective body composed of his assistants and nurses will not be present either.
13. Arthur Koestler talks of the bisociative act in *The Act of Creation* (London: Arkana, 1989).
14. Roger Penrose, interview by the author, June 17, 1998.
15. Tom, interview by the author, June 24, 1998.
16. Christophe Galfard, in *The Hawking Paradox*, directed by William Hicklin, Horizon Films, 2005.
17. His voice irritates, and some participants point out that it would be unbearable.
18. Albert Robillard, *Meaning of a Disability: The Lived Experience of Paralysis* (Philadelphia: Temple University Press, 1999), 52.
19. It is nevertheless true that Hawking sometimes has the entire cosmos in his head, through the use of diagrams.
20. On the materialization of genius and intellectual competencies, see Barthes, *Mythologies*; and Michael Hagner, "The Pantheon of Brains," in *Making Things Public: Atmospheres of Democracy*, ed. Bruno Latour and Peter Weibel (Cambridge, MA: MIT Press, 2005), 126–31; and Steven Shapin, "The Politics of Observation: Cerebral Anatomy and Social Interests in the Edinburgh Phrenology Disputes," in *On the Margins of Science: The Social Construction of Rejected Knowledge*, edited by Roy Wallis (Staffordshire: University of Keele, 1979).
21. William Hicklin, interview by the author, January 31, 2007.
22. His assistants incessantly inscribe what they think about Hawking onto him and the statue.
23. In his case, the projections are sometimes magnified by the computer.
24. See Errol Morris's comments in Arthur Lubow, "Heart and Mind," *Vanity Fair*, June 1992, 47.
25. Michael White and John Gribbin, *Stephen Hawking: A Life in Science* (New York: Dutton, 1992), 292.
26. Tim, interview by the author, June 24, 1998.
27. In numerous movies, and especially in Errol Morris's movie, Hawking is represented as flying with his wheelchair into the cosmos.
28. "According to Weisstein's world of Physics, a graviton is a theoretical particle having no mass and no charge that carries the gravitational force. [...] As of today (June 4, 2003), we have not found proof that the graviton exists. Because these particles carry very little energy, they are very hard to detect—but we're still looking!" Lisa Wei, "Ask an Astronomer," hosted by Cornell University Department of Astronomy (June 2003), available online at http://curious.astro.cornell.edu/question.php?number=535 (accessed May 15, 2008).
29. This recalls Thackery's famous image; see "Rex-Ludovicus-Ludovicus Rex: An Historical Study," in William Makepeace Thackeray, *The Paris Sketch Book, by Mr. Titmarsh* (London: Smith, Elder, 1868), 290.
30. Paul Veyne, *Les Grecs ont-ils crû à leurs mythes? Essai sur l'imagination constituante* (Paris: Poche/Seuil, 2000).

31. On the knowing subject, see also Hélène Mialet, *L'Entreprise créatrice: Le rôle des récits, des objets et de l'acteur dans l'invention* (Paris: Hermès-Lavoisier, 2008).

Conclusion

1. The size and extent of this collectivity depends on where we take the story.
2. Edwin Hutchins, *Cognition in the Wild* (Cambridge, MA: MIT Press, 1995), and "How a Cockpit Remembers Its Speeds," *Cognitive Science* 19 (July–September 1985): 265–88; Charles Goodwin and Marjorie Harness Goodwin, "La coopération au travail dans un aéroport," *Réseaux* 15, no. 85 (1997): 129–62. See also Lucy Suchman, "Constituting Shared Workspaces," in *Cognition and Communication at Work*, ed. Yrjö Engeström and David Middleton (Cambridge: Cambridge University Press, 1996), 35–60.
3. Michel Callon and John Law, "After the Individual in Society: Lessons on Collectivity from Science, Technology and Society," *Canadian Journal of Sociology* 22, no. 2 (1950): 169.
4. Donna Haraway, *Simians, Cyborgs and Women: The Reinvention of Nature* (London: Free Association Books, 1991), 169–81, esp. 177–78.
5. Insofar as the general infrastructure of scientific and technological projects is so deeply based on the premise of individual genius, developing new ways of conceptualizing what genius/innovation is and how it works (through the emergence of what I called "distributed-centered subjects") has substantial implications for the practical organization and recognition of scientific and technological achievement.
6. On the concept of the distributed-centered subject as it operates in the industrial world, see Hélène Mialet, *L'Entreprise créatrice: Le rôle des récits, des objets et de l'acteur dans l'invention* (Paris: Hermès-Lavoisier, 2008).
7. For a comparison between the famous Hawking working in "an invisible" corporation and an unknown applied researcher working in a famous and large corporation (Total), see Hélène Mialet, "Do Angels Have Bodies? Two Stories about Subjectivity in Science: The Cases of William X and Mr. H," *Social Studies of Science* 29, no. 4 (1999): 551–82. For a reflection about the way the distributed-centered subject reincarnates the bodiless knowing subject of the rationalist philosophy and, more surprisingly, of the ANT, see Hélène Mialet, "Reincarnating the Knowing Subject: Scientific Rationality and the Situated Body," *Qui Parle?* 18, no. 1 (2009): 53–73.
8. But again, couldn't one argue that this is the case with many who are disabled? In other words, is it the singularity of Hawking or of his condition? I show, however, on the contrary, the importance of a model of dependence over the model of independence. Thus, while independence was a useful tactical move for the disabled in the 1970s and 1980s, its currency as a primary motivational value is waning. Multiculturalism has made some in the English-speaking world aware of interdependence as an important alternative value. The close examination of Hawking I provide here shatters any illusion of independence toward which many disabled people have striven by showing the reality—and efficacy—of an alternative model: that of the complex system of interdependence and cooperation through which Stephen Haw-

king lives and works. We could also say that we are all, disabled and able alike, situated in complex systems of interdependence and cooperation.

9. I also use the notion of Kantorowicz's two bodies. Indeed I showed how these extended bodies become (at times) naturalized and participate in the construction of his sacred body. See Ernst Kantorowicz, *The King's Two Bodies: A Study in Medieval Political Theology* (Princeton, NJ: Princeton University Press, 1957).
10. These layers or mediations could be infinite. Moreover, substituting the notion of mediation for representation is a way of escaping representationalism. Karen Barad, through her posthumanist notion of performativity, makes a related point: "The idea that beings exist as individuals with inherent attributes, anterior to their representation, is a metaphysical presupposition that underlies the belief in political, linguistic, and epistemological forms of representationalism. Or, to put the point the other way around, representationalism is the belief in the ontological distinction between representations and that which they purport to represent; in particular, that which is represented is held to be independent of all practices of representing." Karen Barad, "Posthumanist Performativity," *Signs: Journal of Women in Culture and Society* 28, no. 3 (2003): 804. She goes on to say, "A posthumanist account calls into question the giveness of the differential categories of 'human' and 'non human,' examining the practices through which these differential boundaries are stabilized and destabilized" (808). Though my ideas are akin to hers, there is a sense in which I reintroduce a form of resistance, or "will," or "a presence," a form of agency concurrent with human intentionality. It is obviously in part a product of attribution, but it is also constructed against and through other beings and things.
11. And by this I don't mean that sometime he is a genius because he is smart, and sometimes he is not.
12. I and the actors along with me (e.g., the journalist, the filmmaker, the sculptor) are working more or less with the same kinds of materials, which is to say, the sociologist-anthropologist does not have privileged access to "the truth."
13. This is the case not only for fieldwork but also for every interaction. Can we ever pretend to know those we are the most familiar with, however close we are to them?
14. For a definition of modernity, see Bruno Latour, *We Have Never Been Modern* (Cambridge, MA: Harvard University Press, 1993).
15. Hawking himself, of course, is included in this "we"; for example, when he argues that to do theoretical physics, one doesn't need more than a good head when, in fact, he relies on the complex apparatus constituted of devices and students that I described.
16. I'm tempted here to use the word *enactment* in the way Annemarie Mol uses it in her beautiful book, *The Body Multiple: Ontology in Medical Practice* (Durham, NC: Duke University Press, 2002). For a definition of agency "as a matter of intra-acting," "as being an enactment and not something that someone or something has," see Barad, "Posthumanist Performativity." To rephrase this question and ask instead to what extent we can exteriorize the competencies of the individual would be to open up the possibility of circling back to singularity through its material and distributed competencies. Each object and person around him can be collectivized, material-

ized, and distributed. To understand why we always return to Hawking, I have tried to show, in each chapter, the work of distribution of competencies (e.g., the roles of the machines and assistants, etc.), but I have also tried to follow the work of singularization (standardization of accounts, role of his body, etc.).

17. Of course stars in science, but also those in show business, art, politics, or industry.
18. The methods and results I have obtained in this study could be used to interrogate not only the scientific personae, but also figures such as the shaman, the saint, and the psychoanalyst. They could also offer possible new ways of explaining the emergence of such emblematic figures as the charismatic leader (Max Weber, *The Theory of Social and Economic Organization* [New York: Free Press, 1964], 358–72); the entrepreneur (Joseph Schumpeter, *Capitalism, Socialism, and Democracy* [New York: Harper Perennial, 2008]); or the artistic genius (Svetlana Alpers, *Rembrandt's Enterprise* [Chicago: University of Chicago Press, 1988]; Nathalie Heinich, *La gloire de Van Gogh* [Paris: Editions de Minuit, 1991]; and Antoine Hennion and Joël-Marie Fauquet, "Authority as Performance," *Poetics* 29 [2001]: 75–88; Antoine Hennion, "La présence de Bach," in *Penser l'oeuvre musicale au XX*[e] *siècle*, ed. Martin Kaltenecker and François Nicolas [Paris: Centre de documentation de la Musique Contemporaine, 2006], 85–94, and "Soli Deo Gloria," *Gradhiva* 12 [2010]: 42–47).
19. One can make here another analogy with Kantorowicz's *The King's Two Bodies*. When the king was buried (entombed), his heart and entrails were separated and put elsewhere—distributed to other important institutions.

BIBLIOGRAPHY

Adler, Jerry, Gerald C. Lubenow, and Maggie Malone. "Reading God's Mind." *Newsweek*, June 13, 1988, 36–39.

Akrich, Madeleine. "The De-scription of Technical Objects." In *Shaping Technology/Building Society: Studies in Sociotechnical Change*, ed. Wiebe Bijker and John Law, 205–24. Cambridge MA: MIT Press, 1992.

Alpers, Svetlana. *Rembrandt's Enterprise: The Studio and the Market*. Chicago: University of Chicago Press, 1988.

Appleyard, Bryan. "A Master of the Universe." *Sunday Times Magazine*, June 19, 1988.

Austin, J. L., *How to Do Things with Words*. New York: Oxford University Press, 1965.

Barad, Karen. "Posthumanist Performativity: Toward an Understanding of How Matter Comes to Matter." *Signs: Journal of Women in Culture and Society* 28, no. 3 (2003): 801–31.

Barnes, Barry. *Scientific Knowledge and Sociological Theory*. London: Routledge and Kegan Paul, 1974.

Barral, Mary Rose. *Merleau-Ponty: The Role of the Body-Subject in Interpersonal Relations*. Pittsburgh, PA: Duquesne University Press, 1965.

Barthes, Roland. *Mythologies*. New York: Hill and Wang, 1987.

Benjamin, Walter. "The Work of Art in the Age of Mechanical Reproduction." In *Illuminations*, 217–51. New York: Schocken Books, 1968.

Biagioli, Mario. *Galileo Courtier: The Practice of Science in the Culture of Absolutism*. Chicago: University of Chicago Press, 1993.

Bloor, David. "Anti-Latour." *Studies in History and Philosophy of Science* 30, no. 1 (1999): 81–112.

———. "Reply to Bruno Latour." *Studies in History and Philosophy of Science* 30, no. 1 (1999): 131–36.

Boslough, John. *Beyond the Black Hole: Stephen Hawking's Universe*. London: Collins, 1985.

———. "Stephen Hawking Probes the Heart of Creation." *Reader's Digest*, February 1984, 39–45.

———. *Stephen Hawking's Universe*. New York: Avon Books, 1985.

———. *Stephen Hawking's Universe: Beyond the Black Holes*. London: Fontana, 1984.

Bourdieu, Pierre. *Science de la science et réflexivité*. Paris: Editions Raisons d'Agir, Collection "Cours et Travaux," 2001.

Bradley, Harriet. "The Seductions of the Archive: Voices Lost and Found." *History of the Human Sciences* 12, no. 2 (1999): 107–22.

Buchler, Justus, ed. *The Philosophy of Peirce: Selected Writings*. London: Routledge, 1940.

Butler, Judith. *Gender Trouble: Feminism and the Subversion of Identity*. London: Routledge, 1990.

Callon, Michel, and Bruno Latour. "Don't Throw the Baby Out with the Bath School! A Reply to Collins and Yearley." In *Science as Practice and Culture*, ed. Andrew Pickering, 343–68. Chicago: University of Chicago Press, 1992.

Callon, Michel, and John Law. "After the Individual in Society: Lessons on Collectivity from Science, Technology and Society." *Canadian Journal of Sociology* 22, no. 2 (1950): 165–83.

———, and John Law. "Agency and the Hybrid Collective." *South Atlantic Quarterly* 94, no. 2 (1995): 481–507.

———, and Vololona Rabeharisoa. "Gino's Lesson on Humanity: Genetics, Mutual Entanglements and the Sociologist's Role." *Economy and Society*, 33, no. 1 (2004): 1–27.

Cantor, Geoffrey. "The Scientist as Hero: Public Images of Michael Faraday." In *Telling Lives in Science: Essays on Scientific Biography*, ed. Richard Yeo and Michael Shortland, 171–95. Cambridge: Cambridge University Press, 1996.

Carruthers, Mary. *The Book of Memory: A Study of Memory in Medieval Culture*. 2nd ed. Cambridge: Cambridge University Press, 2008.

Châtelet, Gilles. *Figuring Space: Philosophy, Mathematics and Physics*. Dordrecht, Holland: Kluwer Academic Publishers, 2000.

Clark, William. "On the Ironic Specimen of the Doctor of Philosophy." *Science in Context* 5, no. 1 (1992): 97–137.

Clifford, James. "Introduction: Partial Truths." In Clifford and Marcus, *Writing Culture*, 1–26.

———, and George Marcus, eds. *Writing Culture: The Poetics and Politics of Ethnography*. Berkeley and Los Angeles: University of California Press, 1986.

Colker, David. "Giving a Voice to the Voiceless." *Los Angeles Times*, May 13, 1997.

Collins, Harry M. *Artificial Experts: Social Knowledge and Intelligent Machines*. Cambridge, MA: MIT Press, 1990.

———. *Changing Order: Replication and Induction in Scientific Practice*. 2nd ed. Chicago: University of Chicago Press, 1992.

———. "The Seven Sexes: A Study in the Sociology of a Phenomenon, or the Replication of Experiment in Physics." *Sociology* 9, no. 2 (1975): 205–24.

———. "The Turing Test and Language Skills." In *Technology in Working Order*, ed. Graham Button, 231–45. London: Sage, 1993.

———, and Steve Yearley. "Epistemological Chicken." In *Science as Practice and Culture*, ed. Andrew Pickering, 301–26. Chicago: University of Chicago Press, 1992.

Cornelis, Gustaaf C. "Analogical Reasoning in Modern Cosmological Thinking." In

Metaphor and Analogy in the Sciences, ed. F. Hallyn, 165–80. Netherlands: Kluwer Academic Publishers, 2000.

Cropper, William H. *Great Physicists: The Life and Times of Leading Physicists from Galileo to Hawking*. Oxford: Oxford University Press, 2001.

Daston, Lorraine. "Condorcet and the Meaning of Enlightenment." *Proceedings of the British Academy* 151 (December 2007): 113–34.

———. "Enlightenment Calculation." *Critical Inquiry* 21, no. 1 (1994): 182–202.

Derrida, Jacques. *Archive Fever: A Freudian Impression*. Trans. Eric Prenowitz. Chicago: University of Chicago Press, 1996.

Descartes, René. *Discours de la méthode pour bien conduire sa raison et chercher la vérité dans les sciences, (plus) la dioptrique-les météores et la géométrie qui sont des essais de cette méthode*. Paris: Fayard, 1987.

———. *The "Meditations" and Selections from the "Principles."* Trans. John Veitch. Chicago: Open Court, 1913.

———. *Méditations métaphysiques*. Paris: Flammarion, 1979.

Dreyfus, Hubert. *What Computers Can't Do: A Critique of Artificial Reason*. New York: Harper & Row, 1972.

———. "Why Computers Must Have Bodies in Order to Be Intelligent." *Review of Metaphysics* 21, no. 1 (1967): 13–32.

Dunning-Davies, Jeremy. "Popular Status and Scientific Influence: Another Angle on 'The Hawking Phenomenon.'" *Public Understanding of Science* 2, no. 1 (1993): 85–86.

Fara, Patricia, *Newton: The Making of Genius*. New York: Columbia University Press, 2002.

Featherstone, Mike. "Archive." *Theory, Culture and Society* 23, nos. 2–3 (2006): 591–96.

Ferguson, Kitty. *Stephen Hawking: Quest for a Theory of Everything*. New York: Bantam Books, 1992.

Filkin, David. *Stephen Hawking's Universe: The Cosmos Explained*. New York: Basic Books, 1998.

Foucault, Michel. "What Is an Author?" In *Language, Counter-Memory, Practice*, ed. Donald F. Bouchard, trans. Donald F. Bouchard and Sherry Simon, 113–38. Ithaca, NY: Cornell University Press, 1977.

Galison, Peter. *Einstein's Clocks, Poincaré's Maps: Empire of Time*. New York: W. W. Norton, 2003.

———. *How Experiments End*. Chicago: University of Chicago Press, 1987.

———. "The Suppressed Drawing: Paul Dirac's Hidden Geometry." *Representations* 72 (Autumn 2000): 145–66.

Garfinkel, Harold. *Studies in Ethnomethodology*. Cambridge: Cambridge University Press: 1967.

Geertz, Clifford. *Works and Lives: The Anthropologist as Author*. Stanford, CA: Stanford University Press, 1988.

———. "Thick Description: Toward an Interpretative Theory of Culture." In *The Interpretation of Cultures*. New York: Basic Books, 1973.

Geison, Gerarld L., and Frederic L. Holmes, eds. "Research Schools, Historical Reappraisals." Special issue, *Osiris* 8 (1993).

Gibbons, Gary, and Stephen Hawking. *Euclidean Quantum Gravity*. Singapore: World Scientific Publishing, 1993.

Gidding, Franklin Henry. *The Principles of Sociology: An Analysis of the Phenomena of Association and Organization*. London: Macmillan, 1896.

Goffman, Erving, *The Presentation of Self in Everyday Life*. Garden City, NY: Doubleday, 1959.

———. *Stigma: Notes on the Management of Spoiled Identity*. Englewood Cliffs, NJ: Prentice-Hall, 1963.

Golledge, Reginald. "On Reassembling One's Life: Overcoming Disability in the Academic Environment." *Environment and Planning D: Society and Space* 15, no. 4 (1997): 391–409.

Goode, David. *A World without Words: The Social Construction of Children Born Deaf and Blind*. Philadelphia, PA: Temple University Press, 1994.

Gooding, David. "What Is Experimental about Thought Experiments?" In *Proceedings of the Biennial Meetings of the Philosophy of Science Association* 2, no. 2 (1992): 280–90.

Goodwin, Charles. "Co-Constructing Meaning in Conversations with an Aphasic Man." *Research on Language and Social Interaction* 28, no. 3 (1995): 233–60.

———. *Conversational Organization: Interaction between Speakers and Hearers*. New York: Academic Press, 1981.

———. "Practices of Seeing, Visual Analysis: An Ethnomethodological Approach." In *Handbook of Visual Analysis*, ed. Theo Van Leeuwen and Carey Jewitt, 157–82. London: Sage, 2000.

———, and Marjorie Harness Goodwin, "La coopération au travail dans un aéroport." *Réseaux* 15, no. 85 (1997): 129–62.

Goody, Jack. *The Domestication of the Savage Mind*. Cambridge: Cambridge University Press, 1977.

Greenblatt, Stephen. *Renaissance Self-Fashioning: From More to Shakespeare*. Chicago: University of Chicago Press, 1980.

Hacking, Ian. *Representing and Intervening: Introductory Topics in the Philosophy of Natural Science* . New York: Cambridge University Press: 1983.

Hagner, Michael. *Cerveaux des génies*. Trans. Olivier Mannoni. Paris: Éditions de la Maison des Sciences de l'Homme, 2008.

———. "The Pantheon of Brains." In *Making Things Public: Atmospheres of Democracy*, ed. Bruno Latour and Peter Weibel, 126–31. Cambridge, MA: MIT Press, 2005.

Haraway, Donna. *Simians, Cyborgs and Women: The Reinvention of Nature*. London: Free Association Books, 1991.

Hawking, Jane. *Music to Move the Stars: A Life with Stephen*. London: Macmillan, 1999.

Hawking, Stephen. *Black Holes and Baby Universes and Other Essays*. New York: Bantam Books, 1993.

———. "A Brief History of a Brief History." *Popular Science*, August 1999, 70–72.

———. *A Brief History of Time*. New York: Bantam Books, 1988.

———, ed. *A Brief History of Time: A Reader's Companion*. Prepared by Gene Stone. New York: Bantam Books, 1992.

———, with Leonard Mlodinow. *A Briefer History of Time*. New York: Bantam Books, 2008.

———. "Imagination and Change: Science in the Next Millennium." VHS tape. Millennium Evenings at the White House.

———. "Playboy Interview: Stephen Hawking-Candid Conversation." *Playboy* 37, no. 4 (April 1990): 63–74.

———. "Prof. Stephen Hawking's Computer Communication System." Online at http://www.hawking.org.uk/index.php/disability/thecomputer.

———. "Prof. Stephen Hawking's Disability Advice." Online at http://www.hawking.org.uk/index.php/disability.

———, ed. *Qui êtes vous Mr Hawking?* Paris: Odile Jacob, 1994.

———. *The Universe in a Nutshell*. New York: Bantam Books, 2001.

Hayles, Katherine. *How We Became Posthuman: Virtual Bodies in Cybernetics, Literature, and Informatics*. Chicago: University of Chicago Press, 1999.

Heidegger, Martin. *Being and Time*. Trans. John Macquarrie and Edward Robinson. New York: Harper & Row, 1962.

Heinich, Nathalie. *La gloire de Van Gogh: Essai d'anthropologie de l'admiration*. Paris: Editions de Minuit, 1991.

Hendriks, Ruud. "Egg Timers, Human Values and the Care of Autistic Youths." *Science, Technology and Human Values* 23, no. 4 (1998): 399–424.

Hennion, Antoine. *La passion musicale*. Paris, Métailié: 2007.

———. "La présence de Bach." In *Penser l'oeuvre musicale au XXe siècle: Avec, sans ou contre l'Histoire?* ed. Martin Kaltenecker and François Nicolas, 85–94. Paris: Centre de documentation de la Musique Contemporaine, 2006.

———. "Soli Deo Gloria. Bach était-il un compositeur." *Gradhiva* 12 (2010): 42–47.

———, and Joël-Marie Fauquet. "Authority as Performance: The Love of Bach in Nineteenth-Century France." *Poetics* 29 (2001): 75–88.

Hicklin, William. *The Hawking Paradox*. Directed by William Hicklin. Horizon Films, 2005.

Hirschauer, Stefan. "The Manufacture of Bodies in Surgery." *Social Studies of Science* 21 (May 1991): 279–319.

Hutchins, Edwin. *Cognition in the Wild*. Cambridge, MA: MIT Press, 1995.

———. "How a Cockpit Remembers Its Speeds." *Cognitive Science* 19, no. 3 (1985): 265–88.

Ihde, Don. *Bodies in Technology*. Minneapolis: University of Minnesota Press, 2002.

———. *Technics and Praxis*. Dordrecht, Holland: D. Reidel, 1979.

Iliffe, Robert. "'Is He Like Other Men?' The Meaning of the *Principia Mathematica*, and the Author as Idol." In *Culture and Society in the Stuart Restoration: Literature, Drama, History*, ed. Gerald MacLean, 159–76. Detroit, MI: Western University Press, 1995.

Jacob, Christian, ed. *Lieux de savoir: Les mains de l'intellect*. Paris: Albin Michel, 2011.

Jarroff, Leon. "Roaming the Cosmos." *Time*, February 8, 1998.

Jones, Matthew. *The Good Life in the Scientific Revolution: Descartes, Pascal, Leibniz, and the Cultivation of Virtue*. Chicago: University of Chicago Press, 2006.

Kaiser, David. *Drawing Theories Apart: The Dispersion of Feynman Diagrams in Postwar Physics*. Chicago: University of Chicago Press, 2005.

———. "Making Tools Travel: Pedagogy and the Transfer of Skills in Postwar Theoretical Physics." In Kaiser, *Pedagogy and Practice of Science*, 41–74.

———, ed. *Pedagogy and Practice of Science: Historical and Contemporary Perspectives*. Cambridge, MA: MIT Press, 2005.

———. "Stick-Figure Realism: Conventions, Reification, and the Persistence of Feynman Diagrams, 1948–1964." *Representations* 70 (Spring 2000): 49–80.

Kampeas, Ron. "Hawking Goes Online with New Computer." *Seattle Times*, March 22, 1997.

Kant, Immanuel. *Critique of Pure Reason*. Trans. and ed. Paul Guyer and Allen Wood. Cambridge: Cambridge University Press, 1999.

Kantorowicz, Ernst H. *The King's Two Bodies: A Study in Medieval Political Theology*. Princeton, NJ: Princeton University Press, 1957.

Keller, Evelyn Fox. *A Feeling for the Organism*. San Francisco: W. H. Freeman, 1983.

———. "The Paradox of Scientific Subjectivity." In Megill, *Rethinking Objectivity*, 313–31.

Kendon, Adam. "Some Functions of Gaze Direction in Social Interaction." *Acta Psychologica* 26, no. 1 (1967): 22–63.

Klein, Ursula, ed. *Tools and Modes of Representation in the Laboratory Sciences*. Dordrecht, Holland: Kluwer Academic Publishers, 2001.

Knorr Cetina, Karin. *Epistemic Cultures: How the Sciences Make Knowledge*. Cambridge, MA: Harvard University Press, 1999.

Knox, Kevin, and Richard Noakes, eds. *From Newton to Hawking: A History of Cambridge University's Lucasian Professors of Mathematics*. Cambridge: Cambridge University Press, 2003.

Koestler, Arthur. *The Act of Creation*. London: Arkana, 1989.

———. *Le cri d'Archimède*. Paris: Calmann-Lévy, 1960.

Koltun, Lilly. "The Promise and the Threat of Digital Options in an Archival Age." *Archivaria* 47 (Spring 1999): 114–35.

Kuhn, Thomas. *The Structure of Scientific Revolutions*. Chicago: University of Chicago Press, 1962.

LaCapra, Dominick. *Writing History, Writing Trauma*. Baltimore, MD: Johns Hopkins University Press, 2001.

Laermans, Rudi, and Pascal Gielen. "The Archive of the Digital An-archive." *Image and Narrative: An Online Magazine of the Visual Narrative* (April 2007). Online at www.imageandnarrative.be/.

Lakoff, Georges. *Women, Fire, and Dangerous Things: What Categories Reveal About the Mind*. Chicago: University of Chicago Press, 1987.

———, and M. Johnson. *Metaphors We Live By*. Chicago: University of Chicago Press, 1979.

Latour, Bruno. *An Actor-Network Theory: A Few Clarifications*. Paper presented at the Center for Social Theory and Technology workshop, Keele University, UK, May 2, 1997.

———. “For David Bloor and Beyond: A Reply to David Bloor’s ‘Anti-Latour.’” *Studies in History and Philosophy of Science* 30, no. 1 (1999): 113–29.

———. *The Pasteurisation of France*. Cambridge, MA: Harvard University Press, 1988.

———. *Science in Action: How to Follow Scientists and Engineers Through Society*. Cambridge, MA: Harvard University Press, 1987.

———. “Sur la pratique des théoriciens.” In *Savoir théoriques et savoirs d’action*, ed. Jean-Marie Barbier. Paris: Presses Universitaires de France, 1998.

———. “A Textbook Case Revisited: Knowledge as a Mode of Existence.” In *The Handbook of Science and Technology Studies*, ed. Edward J. Hackett, Olga Amsterdamska, Michael Lynch, and Judy Wajcman, 83–112. 3rd ed. Cambridge, MA: MIT Press, 2007.

———. “Le travail de l’image ou l’intelligence redistribuée.” *Culture Technique* 22 (1991): 12–24.

———. “Visualization and Cognition: Thinking with Eyes and Hands.” *Knowledge and Society: Studies in the Sociology of Culture Past and Present* 6 (1986): 1–40.

———. *We Have Never Been Modern*. Cambridge, MA: Harvard University Press, 1993.

———. “Which Politics for Which Artifacts?” *Domus*, June 2004, 50–51.

———, and Adam Lowe. “The Migration of the Aura; or, How to Explore the Original Through Its Facsimiles.” In *Switching Codes*, ed. Thomas Bartscherer. Chicago: University of Chicago Press, 2010.

———, and Steve Woolgar. *Laboratory Life: The Social Construction of Scientific Facts*. Beverly Hills, CA: Sage, 1979.

Law, John, and Inguun Moser. “Making Voices: New Media Technologies, Disabilities and Articulation.” In *Digital Media Revisited: Theoretical and Conceptual Innovations in Digital Domains*, ed. Gunnar Liestøl, Andrew Morrison, and Terje Rasmussen, 491–520. Cambridge, MA: MIT Press, 2003.

Lawrence, Christopher, and Steven Shapin, eds. *Science Incarnate: Historical Embodiments of Natural Knowledge*. Chicago: University of Chicago Press, 1998.

Leder, Drew. *The Absent Body*. Chicago: University of Chicago Press, 1990.

Lenoir, Timothy, ed. *Inscribing Science: Scientific Texts and the Materiality of Communication*. Stanford, CA: Stanford University Press, 1998.

Locke, John. *An Essay Concerning Human Understanding*. London: Oxford University Press, 1988.

Lubow, Arthur. “Heart and Mind.” *Vanity Fair*, June 1992, 44–53.

Lupton, Debora, and Wendy Seymour. “Technology, Selfhood and Physical Disability.” *Social Science and Medicine* 50, no. 12 (2000): 1851–62.

Lynch, Michael. “Archives in Formation: Privileged Spaces, Popular Archives and Paper Trails.” *History of the Human Sciences* 12, no. 2 (1999): 65–87.

———. *Art and Artifact in Laboratory Science: A Study of Shop Work and Shop Talk in a Research Laboratory*. London: Routledge and Kegan Paul, 1985.

———. “Discipline and the Material Form of Images: An Analysis of Scientific Visibility.” *Social Studies of Science* 15, no. 1 (1985): 37–66.

———. “The Externalized Retina: Selection and Mathematization in the Visual Documentation of Objects in the Life Sciences.” *Human Studies* 11, nos. 2–3 (1988): 201–34.

———. *Scientific Practice and Ordinary Action: Ethnomethodology and Social Studies of Science*. Cambridge: Cambridge University Press, 1993.

———, and David Bogen, *The Spectacle of History: Speech, Text and Memory at the Iran-Contra Hearings*. Durham, NC: Duke University Press, 1996.

———, Eric Livingston, and Harold Garfinkel. "Temporal Order in Laboratory Work." In *Science Observed*, ed. Karin Knorr Cetina and Michael Mulkay, 205–38. London: Sage, 1983.

———, Stephen Hilgartner, and Carin Berkowitz. "Voting Machinery: Counting and Public Proofs in the 2000 US Presidential Election." In *Making Things Public: Atmospheres of Democracy*, ed. Bruno Latour and Peter Wiebel, 814–28. Cambridge, MA: MIT Press, 2005.

———, and Steve Woolgar, eds. *Representation in Scientific Practice*. Cambridge, MA: MIT Press, 1990.

MacKenzie, Donald. *Knowing Machines: Essays on Technical Change*. Cambridge, MA: MIT Press, 1996.

MacLean, Gerald, ed. *Culture and Society in the Stuart Restoration: Literature, Drama, History*. Detroit, MI: Western University, 1995.

McEvoy, Joseph. P., and Oscar Zarate. *Introducing Stephen Hawking*. New York: Totem Books, 1995.

Megill Allan, ed. *Rethinking Objectivity*. Durham, NC: Duke University Press: 1994.

Merleau-Ponty, Maurice. *Phenomenology of Perception*. Translated by Colin Smith. London: Routledge and Kegan Paul, 1962.

———. *The Structure of Behavior*. Trans. Alden L. Fisher. Boston: Beacon Press, 1963.

Mialet, Hélène. "Do Angels Have Bodies? Two Stories about Subjectivity in Science: The Cases of William X and Mr. H." *Social Studies of Science* 29, no. 4 (1999): 551–82.

———. *L'Entreprise créatrice: Le rôle des récits, des objets et de l'acteur dans l'invention*. Paris: Hermès-Lavoisier, 2008.

———. "Is the End in Sight for the Lucasian Chair? Stephen Hawking as Millennium Professor." In Knox and Noakes, *From Newton to Hawking*, 425–59.

———. "Making a Difference by Becoming the Same." *International Journal of Entrepreneurship and Innovation* 10, no. 4 (2009): 257–65.

———. "Reincarnating the Knowing Subject: Scientific Rationality and the Situated Body." *Qui Parle?* 18, no. 1 (2009): 53–73.

———. "The 'Righteous Wrath' of Pierre Bourdieu." *Social Studies of Science* 33, no. 4 (2003): 613–21.

Michel, Henry. *Généalogie de la psychanalyse*. Paris: Presses Universitaires de France, 1985.

Mol, Annemarie. *The Body Multiple: Ontology in Medical Practice*. Durham, NC: Duke University Press, 2002.

Morfaux, Louis-Marie. *Vocabulaire de la philosophie et des sciences humaines*. Paris: Armand Colin, 1980.

Moser, Ingunn, and John Law. "Good Passages, Bad Passages." In *Actor-Network Theory and After*, ed. John Law and John Hassard, 196–220. Oxford: Blackwell, 1999.

Murphy, Robert F. *The Body Silent*. New York: W. W. Norton, 1987.

Nersessian, Nancy. "The Theoretician's Laboratory: Thought Experimenting as Mental Modeling." *Proceedings of the Biennial Meeting of the Philosophy of Science Association* 2 (1992): 291–301.

Netz, Reviel. *The Shaping of Deduction in Greek Mathematics*. Cambridge: Cambridge University Press, 1999.

Norman, Donald. *Things That Make Us Smart: Defending Human Attributes in the Age of the Machine*. Reading, MA: Addison-Wesley, 1993.

Ochs, Elinor, Sally Jacoby, and Patrick Gonzales. "Interpretive Journeys: How Physicists Talk and Travel Through Graphic Space." *Configurations* 2, no. 1 (1994): 151–71.

Olesko, Kathryn. "The Foundations of a Canon: Kohlrausch's Practical Physics." In Kaiser, *Pedagogy and Practice of Science*, 323–57.

Oudshoorn, Nelly, and Trevor Pinch. *How Users Matter: The Co-Construction of Users and Technology*. Cambridge, MA: MIT Press, 2003.

Penrose, Roger. *The Emperor's New Mind: Concerning Computers, Mind and the Laws of Physics*. New York: Penguin Books, 1989.

Pickering, Andrew. "Against Putting the Phenomena First: The Discovery of the Weak Neutral Current." *Studies in History and Philosophy of Science* 15, no. 2 (1984): 87–117.

———. *The Mangle of Practice: Time, Agency and Science*. Chicago: University of Chicago Press, 1995.

———, and Adam Stephanides. "Constructing Quaternions: On the Analysis of Conceptual Practice." In *Science as Practice and Culture*, ed. Andrew Pickering, 139–67. Chicago: University of Chicago Press, 1992.

Pinch, Trevor. "Giving Birth to New Users: How the Minimoog Was Sold to Rock and Roll." In Oudshoorn and Pinch, *How Users Matter*, 247–70.

———, and Wiebe E. Bijker. "The Social Construction of Facts and Artefacts; or, How the Sociology of Science and the Sociology of Technology Might Benefit Each Other." *Social Studies of Science* 14 (August 1984): 399–441.

Polanyi, Michael. *Knowing and Being*, ed. Marjorie Grene. Chicago: University of Chicago Press, 1969.

———. *The Study of Man: The Lindsay Memorial Lectures*. Chicago: University of Chicago Press, 1959.

———. *The Tacit Dimension*. Garden City, NY: Doubleday,1966.

Popper, Karl. *Conjectures and Refutations: The Growth of Scientific Knowledge*. London: Routledge and Kegan Paul, 1969.

———. *The Logic of Scientific Discovery*. London: Hutchinson, 1972.

———. *Objective Knowledge: An Evolutionary Approach*. Oxford: Clarendon Press, 1972.

Quéré, Louis. "La situation toujours négligée." *Réseaux* 15, no. 85 (1997): 163–92.

Ricoeur, Paul. *Hermeneutics and the Human Sciences: Essays on Language, Action and Interpretation*. Trans. and ed. John B. Thompson. Cambridge: Cambridge University Press, 1981.

———. "The Model of the Text: Meaningful Action Considered as a Text." In Ricoeur, *Hermeneutics and the Human Sciences* , 145–64.

———. "What Is a Text? Explanation and Understanding." In Ricoeur, *Hermeneutics and the Human Sciences* , 197–221.

Ridpath, Ian. "Black Hole Explorer." *New Scientist*, May 4, 1978, 307–9.

Riles, Annelise, ed. *Documents: Artifacts of Knowledge*. Ann Arbor: University of Michigan Press, 2006.

Robillard, Albert. *Meaning of a Disability: The Lived Experience of Paralysis*. Philadelphia, PA: Temple University Press, 1999.

Rosental, Claude. *Weaving Self-Evidence: A Sociology of Logic*. Princeton, NJ: Princeton University Press, 2008.

Rotman, Brian. *Ad Infinitum . . . The Ghost in Turing's Machine: Taking God Out of Mathematics and Putting the Body Back In*. Stanford, CA: Stanford University Press, 1993.

———. "Figuring Figures." Paper presented at an MLA round table conference entitled "Between Semiotics and Geometry: Metaphor, Science, and the Trading Zone." Philadelphia, PA, December 28, 2004.

———. *Mathematics as Sign: Writing, Imagining, Counting*. Stanford, CA: Stanford University Press, 2000.

———. "The Technology of Mathematical Persuasion." In Lenoir, *Inscribing Science*, 55–69.

———. "Thinking Dia-Grams: Mathematics, Writing, and Virtual Reality." *South Atlantic Quarterly* 94, no. 2 (1995): 389–415.

Ryle, Gilbert. "The Thinking of Thoughts: What Is 'Le Penseur' Doing?" In Gilbert Ryle's collected papers, 1971. Online at http://web.utk.edu/~wverplan/ryle.html.

Sacks, Harvey. *Lectures on Conversation*. Vol. 1. Malden: Blackwell, 1992.

———, Emanuel A. Schegloff, and Gail Jefferson. "A Simplest Systematics for the Organization of Turn-Taking for Conversation." *Language* 50, no. 4 (1974): 696–735.

Schaffer, Simon. "Genius in Romantic Natural Philosophy." In *Romanticism and the Sciences*, ed. Andrew Cunningham and Nick Jardine, 82–98. Cambridge: Cambridge University Press, 1990.

———. "Glassworks: Newton's Prism and the Uses of Experiment." In *The Uses of Experiment: Studies in the Natural Sciences*, ed. David Gooding, Trevor J. Pinch, and Simon Schaffer, 67–104. Cambridge: Cambridge University Press, 1989.

———. "Making Up Discovery." In *Dimensions of Creativity*, ed. Margaret Boden, 13–55. Cambridge, MA: MIT Press, 1994.

Schmidt, Sara. *Split Personalities: A Librarian in the Archives*. Special Collections Librarian, Schreiner University, Kerville, Texas.

Schumpeter, Joseph. *Capitalism, Socialism, and Democracy*. New York: Harper Perennial, 2008.

Schweber, Silvan S. *QED and the Men Who Made It: Dyson, Feynman, Schwinger, and Tomonaga*. Princeton, NJ: Princeton University Press, 1994.

Shapin, Steven. "The Mind Is Its Own Place: Science and Solitude in Seventeenth-Century England." *Science in Context* 4 (1991), 191–218.

———. "The Politics of Observation: Cerebral Anatomy and Social Interests in the Edinburgh Phrenology Disputes." In *On the Margins of Science: The Social Construction of Rejected Knowledge*, ed. Roy Wallis. Staffordshire: University of Keele, 1979.

———. *The Scientific Life: A Moral History of a Late Modern Vocation*. Chicago: University of Chicago Press, 2008.

———. *A Social History of Truth: Civility and Science in Seventeenth-Century England*. Chicago: University of Chicago Press, 1994.

———, and Christopher Lawrence. "Introduction." In Lawrence and Shapin, *Science Incarnate*, 1–19.

———, and Simon Schaffer. *Leviathan and the Air-Pump: Hobbes, Boyle, and the Experimental Life*. Princeton, NJ: Princeton University Press, 1989.

Shuttleworth, Russell. "The Pursuit of Sexual Intimacy for Men with Cerebral Palsy." PhD diss., University of California, San Francisco and Berkeley, 2000.

Sibum, Otto. "Les gestes de la mesure: Joule, les pratiques de la brasserie et la science," *Annales Histoire, Sciences Sociales* 4–5 (July–October 1998): 745–74.

———. "Reworking the Mechanical Equivalent of Heat." *Studies in History and Philosophy of Science* 26, no. 1 (1995): 73–106.

Steedman, Carolyn. "Something She Called a Fever: Michelet, Derrida and Dust." *American Historical Review* 106, no. 4 (2001): 1159–80.

Stengers, Isabelle. *Pour en finir avec la tolérance*. Vol. 7 of *Cosmopolitiques*. Paris: Editions La Découverte/Les Empêcheurs de penser en rond, 1997.

Stone, Allucquère Rosanne. "Split Subjects, Not Atoms; or, How I Fell in Love with My Prosthesis." *Configurations* 2, no. 1 (1994): 173–90.

Suchman, Lucy. "Constituting Shared Workspaces." In *Cognition and Communication at Work*, ed. Yrjö Engeström and David Middleton, 35–60. Cambridge: Cambridge University Press, 1996.

———. *Human-Machine Reconfigurations: Plans and Situated Actions*. 2nd ed. Cambridge: Cambridge University Press, 2007.

Tarde, Gabriel. *L'opinion et la foule*. Paris: Presses Universitaires de France, 1989.

Thackeray, William Makepeace. *The Paris Sketchbook, by Mr. Titmarsh*. London: Smith and Elder, 1868.

Thévenot, Laurent. "Pragmatic Regimes Governing the Engagement with the World." In *The Practice Turn in Contemporary Theory*, ed. Theodore R. Schatzki, Karin Knorr Cetina, and Eike von Savigny, 56–74. London: Routledge, 2001.

Thompson, Charis. *Making Parents: The Ontological Choreography of Reproductive Technologies*. Cambridge, MA: MIT Press, 2005.

Thorne, Kip S. *Black Holes and Time Warps: Einstein's Outrageous Legacy*. New York: W. W. Norton, 1994.

Thorpe, Charles, and Steven Shapin. "Who Was J. Robert Oppenheimer? Charisma and Complex Organization." *Social Studies of Science* 30, no. 4 (2000): 545–90.

Turkle, Sherry. *Life on the Screen: Identity at the Age of the Internet*. New York: Touchstone, 1997.

Veyne, Paul. *Les Grecs ont-ils crû à leurs mythes? Essai sur l'imagination constituante*. Paris: Poche/Seuil, 2000.

Warwick, Andrew. *Masters of Theory: Cambridge and the Rise of Mathematical Physics*. Chicago: University of Chicago Press, 2003.

Weber, Max. *The Theory of Social and Economic Organization*. Trans. A. M. Henderson and Talcott Parsons. New York: Free Press, 1964.

Weizenbaum, Joseph. "ELIZA: A Computer Program for the Study of Natural Language Communication between Man and Machine." *Communications of the ACM* 9, no. 1 (1966): 36–45.

White, Hayden. *Metahistory: The Historical Imagination in Nineteenth-Century Europe*. Baltimore, MD: Johns Hopkins University Press, 1973.

White, Michael, and John Gribbin, *Stephen Hawking, A Life in Science*. New York: Dutton, 1992.

Wilford, John Noble. "Scientist at Work: Stephen Hawking— Sailing a Wheelchair to the End of Time." *New York Times*, March 24, 1998.

Wimmer, Mario. *Archivkorper: Eine Geschichte historisher Einbildungskraft*. Constance : Konstanz Univesity Press, 2012.

———. "Die kalte Sprache des Lebendigen: Zu den Anfängen der Archivberufssprache." In *Sprachvollzug im Amt Kommunikation und Verwaltung im Europa des 19 und 20 Jahrhunderts*, ed. Peter Becker. Bielefeld, Germany: Transcript, 2011.

Wintroub, Michael. "The Looking Glass of Facts: Collecting, Rhetoric and Citing the Self in the Experimental Natural Philosophy of Robert Boyle." *History of Science* 35 (June 1997): 189–217.

———. *A Savage Mirror: Power, Identity and Knowledge in Early Modern France*, Stanford, CA: Stanford University Press, 2006.

Yeo, Richard. "Genius, Method, and Morality: Images of Newton in Britain, 1760–1860." *Science in Context* 2, no. 2 (1988): 257–84.

INDEX